T0331169

Sustainable Farming through Machine Learning

This book explores the transformative potential of machine learning (ML) technologies in agriculture. It delves into specific applications, such as crop monitoring, disease detection, and livestock management, demonstrating how artificial intelligence/ machine learning (AI/ML) can optimize resource management and improve overall productivity in farming practices.

Sustainable Farming through Machine Learning: Enhancing Productivity and Efficiency provides an in-depth overview of AI and ML concepts relevant to the agricultural industry. It discusses the challenges faced by the agricultural sector and how AI/ML can address them. The authors highlight the use of AI/ML algorithms for plant disease and pest detection and examine the role of AI/ML in supply chain management and demand forecasting in agriculture. It includes an examination of the integration of AI/ML with agricultural robotics for automation and efficiency. The authors also cover applications in livestock management, including feed formulation and disease detection; they also explore the use of AI/ML for behavior analysis and welfare assessment in livestock. Finally, the authors also explore the ethical and social implications of using such technologies.

This book can be used as a textbook for students in agricultural engineering, precision farming, and smart agriculture. It can also be a reference book for practicing professionals in machine learning and deep learning working on sustainable agriculture applications.

Artificial Intelligence for Sustainable Engineering and Management

Sachi Nandan Mohanty
School of Computer Science & Engineering (SCOPE),
VIT-AP University, Amaravati, Andhra Pradesh, India

Deepak Gupta

Artificial intelligence is shaping the future of humanity across nearly every industry. It is already the main driver of emerging technologies like big data, robotics, and IoT, and it will continue to act as a technological innovator in the foreseeable future. Artificial intelligence is the simulation of human intelligence processes by machines, especially computer systems. Specific applications of AI include expert systems, natural language processing, speech recognition, and machine vision. The future of business intelligence combined with AI will see the analysis of huge quantities of contextual data in real-time. So, the tool will quickly capture customer needs and priorities and do what is needed.

Sustainable Farming through Machine Learning
Edited by Suneeta Satpathy, Bijay Kumar Paikaray, Ming Yang, and Arun Balakrishnan

AI for Climate Change and Environmental Sustainability
Edited by Suneeta Satpathy, Satyasundara Mahapatra, Nidhi Agarwal, and Sachi Nandan Mohanty

Green Metaverse for Greener Economies
Edited by Sukanta Kumar Baral, Richa Goel, Tilottama Singh, and Rakesh Kumar

Healthcare Analytics and Advanced Computational Intelligence
Edited by Sushruta Mishra, Meshal Alharbi, Hrudaya Kumar Tripathy, Biswajit Sahoo, and Ahmed Alkhayyat

AI in Agriculture for Sustainable and Economic Management
Edited by Sirisha Potluri, Suneeta Satpathy, Santi Swarup Basa, and Antonio Zuorro

Deep Learning in Biomedical Signal and Medical Imaging
Edited by Ngangbam Herojit Singh, Utku Kose, and Sarada Prasad Gochhayat

Sustainable Development Using Private AI: Security Models and Applications
Edited by Uma Maheswari V and Rajanikanth Aluvalu

www.routledge.com/AI-for-Sustainable-Engineering-and-Management-series/book-series/AISEM

Sustainable Farming through Machine Learning

Enhancing Productivity and Efficiency

Edited by
**Suneeta Satpathy, Bijay Kumar Paikaray,
Ming Yang, and Arun Balakrishnan**

CRC Press
Taylor & Francis Group
Boca Raton London New York

CRC Press is an imprint of the
Taylor & Francis Group, an **informa** business

ISBN: 978-1-032-77749-8 (hbk)
ISBN: 978-1-032-77750-4 (pbk)
ISBN: 978-1-003-48460-8 (ebk)

DOI: 10.1201/9781003484608

Typeset in Times
by Newgen Publishing UK

Contents

Preface

As the global population continues to grow rapidly, the demand for increased food production has reached unprecedented levels. The challenge is not merely to produce more food but to ensure that farming practices remain sustainable, efficient, and environmentally friendly. The integration of advanced technologies, particularly Artificial Intelligence (AI) and Machine Learning (ML), holds transformative potential to address these challenges directly. This edited volume, "Sustainable Farming through Machine Learning: Enhancing Productivity and Efficiency", examines the convergence of cutting-edge technological advancements and traditional agricultural practices to pave the way for a more sustainable future in farming.

The primary aim of this book is to offer a comprehensive overview of how AI and ML can be leveraged to enhance productivity and efficiency in agriculture. It covers a wide range of topics, from crop health monitoring and disease detection to precision agriculture and decision support systems. Each chapter explores a specific aspect of AI and ML applications in farming, providing insights into the latest research, methodologies, and practical implementations.

Our contributors, a diverse group of experts from academia, industry, and the field, bring a wealth of knowledge and experience to this endeavour. Their collective efforts highlight innovative approaches and solutions that are already making significant impacts across various agricultural landscapes worldwide. From the use of drones for crop monitoring to the development of AI-driven virtual tutors for farmer education, the chapters in this book demonstrate the myriad ways technology is revolutionizing farming practices.

This book is structured to guide readers through a logical progression of topics, beginning with foundational strategies for crop health monitoring and management and advancing to more complex systems like autonomous sensor networks and ethical considerations. We have aimed to cover both theoretical frameworks and practical case studies, ensuring that the content is accessible to a broad audience, including researchers, practitioners, policymakers, and students.

As we navigate through the contents of this book, we hope to inspire new ideas, foster collaborations, and drive further research and development in the realm of sustainable farming. The fusion of AI and ML with agriculture holds immense promise, not only for boosting productivity but also for safeguarding our environment and promoting the well-being of farming communities.

We would like to extend our gratitude to all the contributors for their invaluable insights and to the readers who are taking the time to engage with this important subject. Together, we can work towards a future where agriculture is not only more productive and efficient but also more resilient and sustainable.

Editors

About the editors

Suneeta Satpathy, PhD, is an Associate Professor in the Center for AI & ML, Siksha 'O' Anusandhan (Deemed to be) University, Odisha, India. Her research interests include computer forensics, cyber security, data fusion, data mining, big data analysis, decision mining, and machine learning. She has published papers in many international journals and conferences in repute. She has two Indian patents to her credit and is a member of IEEE, CSI, ISTE, OITS, and IE.

Bijay Kumar Paikaray, PhD, is an Associate Professor at the Center for Data Science, Siksha 'O' Anusandhan (Deemed to be) University, Odisha. His interests include high-performance computing, information security, machine learning, and IoT.

Ming Yang has a PhD in Computer Science from Wright State University, Dayton, Ohio, US, 2006. Currently he is a Professor in the College of Computing and Software Engineering Kennesaw State University, GA, USA. His research interests include multimedia communication, digital image/video processing, computer vision, and machine learning.

Arunkumar Balakrishnan, PhD, holds the position of Assistant Professor Senior Grade in the Computer Science and Engineering department at VIT-AP University. He obtained his PhD in Information Science and Engineering from Anna University, Chennai. He possesses 12 years of academic expertise and an additional 6 years of concurrent research experience in the domains of Cryptography, Medical Image Security, Blockchain, and NFT. His research interests encompass Cryptography, Network Security, Medical Image Encryption, Blockchain, lightweight cryptography methods, and NFT.

Contributors

Helensharmila A.
VIT-AP University, Amaravati,
  Andhra Pradesh, India

Chintala Sai Akshitha
Department of Computer Science and
  Engineering Chaitanya Bharathi
  Institute of Technology(A), Gandipet,
  Hyderabad, India

T. Anithakumari
School of Computer Science and
  Engineering, VITAP University,
  Amaravati, Andhra Pradesh

Priya Banerjee
Department of Computer Science and
  Engineering,
Centurion University of Technology and
  Management, Odisha, India

Swadhin Kumar Barisal
Department of Computer Application
ITER, FET, Siksha 'O' Anusandhan
  (Deemed to be University),
Bhubaneswar, India

Sujata Chakravarty
Department of Computer Science and
  Engineering,
Centurion University of Technology and
  Management, Odisha, India

Santilata Champati
Department of Mathematics
ITER, FET, Siksha 'O' Anusandhan
  deemed to be University
  Bhubaneswar, India

Sung-Bae Cho
Department of Computer Science,
  Yonsei University, Sudaemoon-gu,
  Seoul, South Korea

Abishi Chowdhury
School of Computer Science
  and Engineering, Vellore
  Institute of Technology,
  Chennai, Tamil Nadu,
  India

Somu Preethi Deekshita
School of Computer Science &
  Engineering (SCOPE), VIT-AP
  University, Amaravati,
  Andhra Pradesh, India

Jeethu V. Devasia
School of Computer Science and
  Engineering, VIT-AP University,
  Amaravati, Andhra Pradesh, India

Smitta Ranjan Dutta
Ajay Binay Institute of
  Technology (ABIT), Cuttack,
  Odisha, India

Mamata Garanayak
Department of Computer Science,
  Kalinga Institute of Social
  Sciences, Deemed to be University,
  Bhubaneswar, Odisha, India

Yaddanapudi Renuka Gayathri
School of Computer Science &
  Engineering (SCOPE), VIT-AP
  University, Amaravati,
  Andhra Pradesh, India

Suwarna Gothane
Department of Computer
  Engineering, Pimpri
  Chinchwad College of Engineering
  Pune, Maharashtra, India

Ojasva Jain
School of Computer Science and
  Engineering, VITAP University,
  Amaravati, Andhra Pradesh, India

Arunima Jaiswal
Indira Gandhi Delhi Technical
  University for Women (IGDTUW),
  Delhi, India

Lambodar Jena
Department of Computer Science and
  Engineering, Siksha 'O' Anusandhan
  (Deemed to be) University,
  Bhubaneswar, Odisha, India

Swarna Prabha Jena
Department of Electronics and
  Communication Engineering,
Centurion University of Technology and
  Management, Odisha, India

Ramireddy Jyothsna
School of Computer Science &
  Engineering (SCOPE), VIT-AP
  University, Amaravati,
  Andhra Pradesh, India

P. Venkata Kishore
CSM Department, St. Peters
  Engineering College, Hyderabad,
  Telangana State, India

Pushkar Kishore
Department of Computer Science &
  Engineering National Institute of
  Technology Rourkela, Odisha, India

M. Dilip Kumar
CSE Department, St. Peters Engineering
  College, Hyderabad, Telangana State,
  India

D. Naveen Kumar
EEE Department, Guru Nanak
  Institutions Technical Campus,
  Hyderabad, Telangana State, India

M. Vijay Kumar
Department of AI&DS
B V Raju Institute of Technology
  Narsapur, Telangana State, India

S. Vijay Kumar
EEE Department, Dr. KV SubbaReddy
  Engineering College, Hyderabad,
  Telangana State, India

Suprava Ranjan Laha
Department of Computer Science and
  Engineering, Institute of Technical
  Education and Research, Siksha
  'O' Anusandhan (Deemed to be
  University), Bhubaneswar, India

Mohamed Iqbal M.
Department of Information Technology,
  Sri Sivasubramania Nadar College of
  Engineering, Chennai, Tamil Nadu,
  India

Venkata Krishna Reddy M.
Department of Computer Science and
  Engineering Chaitanya Bharathi
  Institute of Technology(A) Gandipet,
  Hyderabad, India

G. Mamatha
Department of Computer Engineering
and Technology, Chaitanya Bharathi
Institute of Technology(A), Gandipet,
Hyderabad, and Research Scholar,
Department of Computer Science
and Engineering, B.M.S College of
Engineering, Bengaluru, Karnataka,
India

Smita Maurya
Indira Gandhi Delhi Technical
University for Women (IGDTUW),
Delhi, India

Sanket Mishra
School of Computer Science and
Engineering, VITAP University,
Amaravati, Andhra Pradesh

Shruti Mishra
Centre for Advanced Data Science,
Vellore Institute of Technology,
Chennai, Tamil Nadu, India

Soumya Priyadarshini Mishra
Research Scholar, Biju Patnaik
University of Technology (BPUT),
Rourkela, Odisha, India

Sushruta Mishra
School of Computer Engineering,
Kalinga Institute of Industrial
Technology (Deemed to be)
University, Bhubaneswar,
Odisha, India

Bijayini Mohanty
Center for Data science
ITER, Siksha 'O' Anusandhan (Deemed
to be) University
Bhubaneswar, India

Munmun Mohanty
Global Institute of Management,
Bhubaneswar, Odisha,
India

Sachi Nandan Mohanty
School of Computer Science &
Engineering (SCOPE), VIT-AP
University, Amaravati, Andhra
Pradesh, India

Umarani Nagavelli
Department of AI&DS
B V Raju Institute of Technology
Narsapur, Telangana State, India

Preethi Nanjundan
Department of Data Science
CHRIST University, Pune Lavasa
Campus, Lavasa, Maharashtra, India

Gayatri Nayak
Dept. of Comp. Sc. & Engineering
ITER, FET, Siksha 'O' Anusandhan,
(Deemed to be) University,
Bhubaneswar, India

Jyothi S. Nayak
Department of Computer Science and
Engineering, B.M.S College of
Engineering, Bengaluru, Karnataka

Soumen Nayak
Department of Computer Science and
Engineering, Siksha 'O' Anusandhan
(Deemed to be) University,
Bhubaneswar, Odisha, India

Nandini Nenavath
Department of Computer Science and
Engineering Chaitanya Bharathi
Institute of Technology (A),
Gandipet, Hyderabad, India

Fatimun Nisha
Department of Computer Science and
  Engineering,
Centurion University of Technology and
  Management, Odisha, India

Deepanramkumar P.
VIT-AP University, Amaravati,
  Andhra Pradesh, India

Bijay Kumar Paikaray
Centre for Data Science,
Department of Computer Science and
  Engineering,
Siksha 'O' Anusandhan (Deemed to be
  University) University, Odisha, India

Pranati Palai
Department of Teacher Education,
  RNIASE, Cuttack, Odisha, India

Archisman Panda
School of Computer Science and
  Engineering, Vellore Institute of
  Technology, Chennai, Tamil Nadu,
  India

Arjav Anil Patel
School of Computer Science and
  Engineering, Vellore Institute of
  Technology, Chennai, Tamil Nadu,
  India

Kumar Janardan Patra
Department of Computer Science and
  Engineering, Odisha University of
  Technology and Research, Odisha,
  Bhubaneswar, India

Ramesh Patra
Department of Computer Science,
  Kalinga Institute of Social
  Sciences, Deemed to be University,
  Bhubaneswar, Odisha, India

Binod Kumar Pattanayak
Department of Computer Science and
  Engineering, Institute of Technical
  Education and Research, Siksha
  'O' Anusandhan (Deemed to be)
  University, Bhubaneswar, India

Indu P.V.
Department of Data Science
CHRIST University, Pune Lavasa
  Campus, Lavasa, Maharashtra, India

Swayumjit Ray
Ajay Binay Degree College (ABIT),
  Cuttack, Odisha, India

K.V.B. Reddy
EEE Dept., St. Peters Engineering
  College, Hyderabad, Telangana State,
  India

Nitin Sachdeva
Amity University, Noida

Amaresh Sahu
Ajay Binay Institute of Technology,
  Master in Computer Application,
Cuttack, Odisha, India

Sandeep Kumar Satapathy
Department of Computer Science,
  Yonsei University, Sudaemoon-gu,
  Seoul, South Korea

Chinmayee Senapati
Department of Civil Engineering,
  Institute of Technical Education and
  Research, Siksha 'O' Anusandhan
  (Deemed to be) University,
  Bhubaneswar, India

Swagatika Senapati
Department of Civil Engineering,
Institute of Technical Education
and Research, Siksha 'O'
Anusandhan (Deemed to be)
University, Bhubaneswar,
India

Muhammad Zakaria Shaik
School of Computer Science and
Engineering, Vellore Institute of
Technology, Chennai, Tamil Nadu,
India

Kanchan Sharma
Department of Electronics &
Communications Engineering, Indira
Gandhi Delhi Technical University
for Women, New Delhi, Delhi, India

T. Sreeja
Department of AI&DS
B V Raju Institute of Technology
Narsapur, Telangana State, India

K. Sree Latha
EEE Department, St. Peters Engineering
College, Hyderabad, Telangana State,
India

Manas Kumar Swain
Department of Computer Science
and Engineering, Bhubaneswar
Engineering College, Bhubaneswar,
Odisha, India

Sanata Kumar Swain
Ajay Binay Institute of Technology
(ABIT), Cuttack, Odisha, India

Satyaprakash Swain
Department of Computer Science and
Engineering, Institute of Technical
Education and Research, Siksha
'O' Anusandhan (Deemed to be)
University, Bhubaneswar, India

Lijo Thomas
Department of Psychology
CHRIST University, Pune Lavasa
Campus, Lavasa, Maharashtra, India

Alakananda Tripathy
Center for AI & ML
ITER, FET, Siksha 'O' Anusandhan
(Deemed to be) University
Bhubaneswar, India

Srilakshmi V.
Department of CSE (AI&ML) B V Raju
Institute of Technology
Narsapur, Telangana State, India

Gokul Yenduri
VIT-AP University, Amaravati,
Andhra Pradesh, India

1 Exploring AI and ML Strategies for Crop Health Monitoring and Management

Preethi Nanjundan, Indu P.V., and Lijo Thomas

INTRODUCTION

Life depends on agriculture, yet it confronts many obstacles that are made worse by population increase, climate change, and environmental deterioration [1]. Improving agricultural methods while reducing their negative environmental effects [2] is imperative given the growing need for food supply. This chapter examines the revolutionary solutions that artificial intelligence (AI) and machine learning (ML) technology provide, specifically in the field of agricultural health monitoring. Modern technology relies heavily on AI and ML to enable computers to mimic human intelligence [3]. These technologies allow systems to evaluate large datasets, identify patterns, and make well-informed judgments on their own. These technologies have the ability to completely transform crop health monitoring in agriculture by evaluating a wide range of data sources, including sensor data, satellite imaging, and historical records. Through the early detection of crop disease indications, nutrient AI/ML enables prompt interventions to address inadequacies and pest infestations, maximizing yields and resource use. This chapter will explore many AI/ML approaches designed for crop health monitoring, such as deep learning architectures, reinforcement learning strategies, and supervised and unsupervised learning algorithms. Additionally, we will go over useful strategies for implementing AI/ML models in agricultural contexts, such as engineering, data pre-treatment, and model assessment. Furthermore, ethical issues and responsible deployment techniques will be highlighted, emphasizing the necessity of openness, responsibility, and equity in AI-powered agricultural systems. The ultimate goal of this chapter is to picture a time when intelligent technology will be essential in developing resilient and sustainable agricultural systems that will guarantee food security for future generations [4].

DOI: 10.1201/9781003484608-1

1.1 INTRODUCTION TO AI/ML TECHNIQUES FOR CROP HEALTH MONITORING

Monitoring crop health is essential to maintaining food security and sustainable agriculture, especially in light of the growing difficulties brought on by population expansion, climate change, and environmental degradation. Techniques for ML and AI have become more potent tools for improving agricultural health monitoring in recent years. An overview of AI/ML [5] approaches designed especially for crop health monitoring is given in this chapter, with an emphasis on how they may transform farming methods and reduce production risks. Thanks to AI and ML technology, agricultural systems can now make use of enormous volumes of data from various sources, such as sensor networks, satellite photos, and historical records [6]. These technologies use sophisticated algorithms to scan large datasets and find patterns, anomalies, and very accurate crop health indicator predictions [7]. With the use of this capacity, farmers and other agricultural stakeholders may optimize resource allocation, pest control tactics, and crop treatment interventions in real-time by making data-driven decisions. This chapter will examine some AI/ML methods, such as supervised, unsupervised, and deep learning approaches, often used in crop health monitoring. For example, supervised learning algorithms make it possible to classify crop diseases using labeled training data [8,9], whereas unsupervised learning approaches make it easier to find anomalies and do clustering analysis on big datasets. Convolutional neural networks (CNNs) are one type of deep learning architecture that is particularly good at extracting complex characteristics from pictures and time-series data. This allows for the accurate diagnosis of crop diseases and stresses. Additionally, we will review practical issues with using AI/ML approaches for crop health monitoring, such as feature selection, data preparation, and model validation. Furthermore, a discussion will be provided on ethical issues and responsible deployment techniques, highlighting the significance of responsibility, justice, and transparency in AI-powered agricultural systems [10]. Crop health monitoring may be made more proactive and data-driven by utilizing AI and ML, which will help farmers maximize crop yields, reduce their environmental impact, and guarantee sustainable food production for the world's expanding population.

1.1.1 SUPERVISED LEARNING AND UNSUPERVISED LEARNING IN CROP HEALTH MONITORING

Crop health monitoring relies heavily on two core ML techniques: supervised learning and unsupervised learning. The system is trained on labeled data in supervised learning, where each data point is connected to a particular class or category. This method works especially well for applications like yield prediction, pest detection, and disease categorization in agricultural health monitoring. For instance, by examining characteristics taken from photos or sensor data, supervised learning systems may be trained to distinguish between disease-free plants and those that are afflicted [11]. The system can effectively classify new cases and offer insightful information about crop health status since it was trained on a varied dataset that included labeled samples of healthy and ill plants. Conversely, however, [12] unsupervised learning looks for hidden patterns and structures in the data rather than depending on labeled input. In crop health monitoring, this method works well for tasks including anomaly identification, clustering analysis, and pattern recognition. In agricultural photography or sensor data, for example, unsupervised learning algorithms might spot odd patterns or outliers that point to possible problem areas like insect infestations or nutrient shortages. In addition, comparable occurrences of crop health indicators may be grouped together [13] using unsupervised learning techniques like clustering. This allows for focused treatments and offers important insights into the underlying structure of the data; both supervised and unsupervised learning approaches have special benefits in the context of crop health monitoring, and they are frequently combined to produce thorough insights about crop state of health. While unsupervised learning offers useful exploratory analysis and anomaly detection skills, supervised learning allows for accurate classification and prediction jobs. Agricultural stakeholders may improve resource allocation, monitor crop health [7], and reduce production risks by utilizing the capabilities of both techniques. This will eventually promote sustainable agriculture and food security.

1.1.2 Reinforcement Learning: Learning Through Interaction and Feedback in Agriculture

Given its capacity to learn from interactions with the environment and the feedback obtained from those interactions, reinforcement learning (RL), an ML paradigm, has great potential to improve agricultural methods. Recursive logic models may be trained to make agricultural decisions that maximize long-term benefits, such as crop output, resource efficiency, and environmental sustainability [14]. Fundamentally, RL entails an agent that moves through an environment and picks up cues from its surroundings, such as rewards or punishments. An autonomous vehicle, robotic device, or decision-making algorithm might be the agent in an agricultural scenario, while the environment could be a field, greenhouse, or farming business. The agent discovers which activities result in positive results by interacting with the environment, and it modifies its conduct throughout time in line with that. Autonomous crop management is a major usage of RL in agriculture, where RL algorithms are applied to optimize fertilization, irrigation, and pest management techniques. For instance, an RL agent placed in a greenhouse may be trained to modify watering schedules in response to real-time sensor data, increasing crop growth and reducing water use [15]. Similarly, robotic devices or drones may be used most effectively for precision farming operations like planting, harvesting, and weed management using RL models. Personalized farming suggestions, where computers learn to customize agricultural techniques to particular environmental circumstances and crop requirements, is another area where RL shows potential. RL models may adjust agricultural practices by continually learning from feedback obtained from yield data and field observations and adapting to shifting circumstances, resulting in increased sustainability and productivity [16]. All things considered, RL provides an effective framework for autonomous decision-making and adaptive management techniques, which have the potential to revolutionize agriculture. Real-time (RL) models have the potential to improve food production systems' efficiency, resilience, and sustainability by optimizing resource allocation, enhancing crop output, and promoting environmental conservation through inter-action and feedback.

1.1.3 IMPORTANCE OF QUALITY DATA AND DATA PRE-TREATMENT IN AGRICULTURAL APPLICATIONS

In agricultural applications, high-quality data and efficient data pre-treatment are critical to the accuracy and dependability of ML models and the farming methods they subsequently influence [17]. Ensuring the integrity of data inputs is critical for gaining useful insights and making choices in the agricultural sector, where environ-mental conditions can vary greatly and crop health is impacted by multiple factors. First and foremost, ML algorithms are built on top of high-quality data. In the context of agriculture, this means gathering a variety of datasets that cover factors, including crop health indicators, weather patterns, soil composition, and historical yield data. Through the utilization of superior data sources such as satellite images, Internet of Things sensors [18], and remote sensing technologies, scholars and farmers may acquire a thorough comprehension of agricultural ecosystems and recognize patterns and trends that assist in making decisions. However, noise, outliers, and missing values are frequently present in raw data gathered from agricultural settings, which might impair the effectiveness of ML models. Thus, datasets are cleaned, pre-processed, and standardized using data pre-treatment techniques before input into algorithms for analysis. Making sure the input data is appropriate for training the model entails doing things like eliminating superfluous features, imputing missing values, normalizing data distributions, and correcting class imbalances. Additionally, pre-treatment of the data is essential for reducing biases and enhancing the gener-alizability of ML models in agricultural contexts [2]. Researchers may reduce the effect of confounding factors and ensure that models are reliable and flexible across many contexts by carefully selecting and preparing datasets, agricultural settings, and circumstances. Ultimately, it is impossible to overestimate the significance of high-quality data and data pre-treatment in agricultural applications. Stakeholders in the agricultural sector can fully realize the potential of ML technologies to optimize crop production, enhance resource management procedures, and support sustainable farming initiatives by utilizing precise and dependable data inputs and implementing efficient pre-processing techniques [19].

1.2 NEURAL NETWORKS AND DEEP LEARNING FOR CROP HEALTH MONITORING

Deep learning methods and neural networks have become highly effective tools for crop health monitoring because they provide sophisticated features for feature extraction, pattern identification, and predictive analytics. These technologies are essential for automating and improving monitoring operations in the agricultural sector, where prompt diagnosis and control of crop diseases and stressors are cru-cial for optimizing yields and guaranteeing food security. Neural networks are com-puter models [20] made up of linked nodes, or neurons, arranged in layers. They are inspired by the structure and operation of the human brain. Large datasets of labeled samples are used to train these networks to identify pertinent characteristics and forecast outcomes based on input data. Neural networks are used in crop health monitoring to interpret different agricultural data formats, such as multispectral imaging, meteorological data, and soil properties, to spot trends that could point to insect infestations, nutrient shortages, or crop illnesses. Using many layers of linked neurons to extract hierarchical representations of input, deep learning,[7] a subset of neural network methods, expands the capabilities of conventional ML models. Among the deep learning architectures frequently used in crop health monitoring applications are generative adversarial networks (GANs), recurrent neural networks (RNNs), and CNNs. For example, CNNs are very good at processing visual data and have been used successfully for plant disease diagnosis and leaf segmentation. CNNs can precisely identify and locate unhealthy areas within crops by learning hierarchical characteristics from raw picture inputs. This allows for early identification and focused treatments. Alternatively, RNNs are ideal for examining sequential data, including crop health indicator time-series observations. By modeling temporal dependencies and dynamics, these networks make it easier to predict crop phenology, estimate yields, and plan the best irrigation practices. Moreover, GANs make it possible to create synthetic data, which may be utilized to supplement small training datasets and strengthen the generalizability and robustness of models for crop health monitoring. In conclusion, neural networks and deep learning methods provide adaptable and powerful solutions for crop health monitoring, enabling researchers and farmers to track crop conditions, identify irregularities, and maximize management strategies with previously unheard-of precision and effectiveness. These technologies have a lot of potential to advance agricultural sustainability and resistance to new environmental issues as they develop [21].

1.2.1 Simulating Human Brain With Neural Networks in Agriculture

Neural network simulation of the complex operations of the human brain has transformed several sectors, including agriculture. Neural networks, which imitate the brain's information processing and data-driven learning capabilities, provide novel approaches to complicated agricultural problems. Neural networks are used in agriculture to assess large volumes of data gathered from several sources, including weather stations, soil sensors, and remote sensing [22]. These networks are made up of linked nodes, or neurons, arranged in layers that resemble the architecture of the human brain. Neural networks may be trained to see patterns and correlations in data, allowing them to forecast and make judgments. Crop yield prediction is one of the main uses of neural networks in agriculture. The examination of past data on neural networks has the ability to accurately predict future crop yields based on several parameters, including weather patterns, soil composition, and crop management techniques [23]. Farmers may improve yield while limiting resource inputs by using this information to optimize planting timings, irrigation techniques, and fertilizer applications. Neural networks are also essential for detecting pests and diseases in crops. Neural networks can recognize minute visual signals that indicate pest infestations or disease signs by examining photographs taken by drones or satellites. Farmers may reduce crop losses and slow the spread of pests and diseases by using proactive measures like crop rotations and tailored pesticide treatments, which are made possible by early detection. Neural networks are also utilized in precision agriculture to maximize resource utilization. Through the integration of data, neural networks can produce exact suggestions for irrigation, fertilization, and pesticide application customized to each crop's individual requirements based on sensors [20] that track soil moisture, nutrient levels, and crop health [10]. By using fewer inputs, this focused strategy minimizes the environmental impact while increasing production and quality.

1.2.2   DEEP LEARNING: MULTI-LAYERED NEURAL NETWORKS FOR CROP HEALTH MONITORING

A kind of ML called deep learning has become a potent tool for crop health monitoring because of its unmatched capacity to analyze intricate agricultural data. Multi-layered neural networks, the foundation of deep learning, allow us to extract complex patterns and characteristics from huge datasets, transforming crop health monitoring and management. Deep neural networks, or multi-layered neural networks, comprise several layers of linked neurons that analyze incoming data by applying successive transformations. Because these networks can learn [21] hierarchical representations of data, they can comprehend the complex linkages and subtleties seen in datasets related to agriculture. Deep learning models are particularly good at tasks like object recognition, anomaly detection, and picture classification in crop health monitoring. As an illustration, deep CNNs are frequently employed to evaluate remotely sensed crop photos and spot disease, stress, or nutrient deficiency indicators. CNNs can learn to discriminate between healthy and unhealthy plants with exceptional accuracy by using massive, annotated datasets for training, which makes early diagnosis and intervention possible. Moreover, time-series data analysis is done using RNNs and long short-term memory networks (LSTMs) to examine things like crop growth trends, meteorological conditions, and insect invasion dynamics [23]. By capturing temporal relationships and variations in data, these networks make it possible to identify patterns or abnormalities that may need to be addressed and anticipate future crop health status. Additionally, research is being done on the use of GANs to create artificial agricultural data, enhance already-existing datasets, enable reliable model training, and enhance the performance of generalization. Moreover, realistic simulations of agricultural development and environmental factors may be produced using GANs, allowing researchers to investigate different situations and improve management techniques. In conclusion, multi-layered neural networks, which underpin deep learning techniques, have many potential uses in crop health monitoring. Researchers and practitioners may improve decision-making, get a better knowledge of agricultural systems, and ultimately contribute to the sustainability and resilience of food production by utilizing these models' ability to learn intricate patterns and representations from data. Deep learning is expected to become increasingly important in determining how agriculture develops in the future [24].

1.2.3 APPLICATIONS OF DEEP LEARNING IN CROP DISEASE DETECTION AND IMAGE RECOGNITION

The power of deep learning to automatically extract hierarchical representations from data has transformed agricultural picture identification and crop disease diagnosis. Deep learning applications in these fields have greatly increased disease identification and diagnosis efficiency and accuracy, improving crop management techniques. CNNs, one type of deep learning model, have shown remarkable effectiveness in the field of crop disease identification. Large crop picture datasets may be analyzed by these algorithms, which can automatically identify pertinent characteristics and patterns linked to various illnesses. For example, a trained CNN may diagnose patients quickly and accurately by differentiating between healthy plants and those that are sick with particular infections. Crop health monitoring relies heavily on image recognition, identifying visual clues linked to illnesses, dietary deficits, or vermin infestations. Using their ability to recognize minute details and minor fluctuations in pictures, deep learning models do exceptionally well in this challenge. Consequently, these models are able to precisely identify crop-related problems by classifying and labeling photos according to particular attributes. Deep learning models [25] can recognize various plant species, development phases, and environmental conditions in general picture identification tasks, which goes beyond disease diagnosis. When it comes to making educated decisions about irrigation, fertilization, and pest management, this information is crucial for assessing the general health of crops. Deep learning aids in sustainable agricultural operations by automating and enhancing the precision of disease diagnosis and picture identification. Researchers and farmers can quickly recognize and address new problems, such as reducing crops. Practical Considerations in AI/ML Techniques for Crop Health Monitoring losses and maximizing the usage of resources. Deep learning applications in agriculture have a lot of potential to improve crop health monitoring and guarantee global food security as the area develops.

1.3 PRACTICAL CONSIDERATIONS IN AI/ML TECHNIQUES FOR CROP HEALTH MONITORING

To ensure efficacy and dependability in actual agricultural settings, several practical considerations must be carefully considered when using artificial intelligence and machine learning (AI/ML) approaches for crop health monitoring. First off, creating AI/ML models for crop health monitoring requires careful consideration of feature engineering. It entails picking [25] and removing essential aspects from unprocessed data, including spectral reflectance readings from distant sensors or picture features. The model's capacity to identify significant patterns connected to crop health indicators, such as disease symptoms or nutrient deficits, is improved by effective feature engineering. An additional crucial component of practical concerns in AI/ML approaches for crop health monitoring is model assessment. It is possible to evaluate the effectiveness of trained models by utilizing suitable measures, such as accuracy, precision, recall and F1 score [26], and efficiency in locating and evaluating crop health problems. Additionally, validation methods like holdout validation and cross-validation shed light on the model's possible limits and ability to generalize.

Developing strong AI/ML models for crop health monitoring requires striking [27] a balance between bias and variation. Bias is the model's propensity to oversimplify intricate connections between input characteristics and target variables, whereas variance is the model's sensitivity to changes in training data. The model may reliably reflect underlying trends in various agricultural situations without overfitting or underfitting if the variance and bias are balanced appropriately. In addition, resolving issues with data quality and pre-processing are practical factors in AI/ML systems for crop health monitoring, ensuring the availability of noise-free, bias-free training data of the highest caliber. In order to train dependable models, missing values are necessary. The quality and diversity of training data are improved by data preparation techniques, including normalization, feature scaling, and data augmentation, improving the model's performance and capacity for generalization.

1.3.1 Feature Engineering for Agriculture

Utilizing AI/ML approaches for agricultural crop health monitoring requires feature engineering. Choosing and finding pertinent agricultural metrics or indicators that may precisely reflect the health and condition of crops is what feature engineering entails in this context. These factors could include soil properties like moisture [28] content and nutrient levels, vegetation indices produced from satellite or drone photography, meteorological variables like temperature and humidity, and historical crop production data. To find the most useful features for crop health monitoring, feature engineering requires subject expertise. Engineers can choose characteristics that capture the finer points of crop health dynamics by utilizing agricultural expertise, sensor technology, and data analytics. Furthermore, transformation methods like normalization, scaling, and dimensionality may be used in feature engineering. reduction to enhance the chosen characteristics' quality and interpretability [29]. AI/ML models may successfully discover patterns and correlations within the data by carefully developing elements important to agricultural settings. This can result in crop health monitoring systems that are more reliable and accurate. Moreover, continuous research and advancements in feature engineering techniques further augment the capacities of AI/ML approaches in tackling the intricate problems of food security and sustainable agriculture.

1.3.2 Model Assessment in Crop Health Monitoring

A crucial part of AI/ML methods for crop health monitoring is model assessment, which guarantees the precision and dependability of predictive models used in farming applications. In order to determine how well ML algorithms forecast crop health indicators, they must analyze their performance using a variety of measures and methodologies. Evaluation criteria, including accuracy, precision, recall, F1 score, and area under the receiver operating characteristic curve (AUC-ROC) [30], are frequently used in model assessment. These metrics shed light on how well the model uses input characteristics to properly categorize crops as healthy or unhealthy. Furthermore, methods like cross-validation, which divides the dataset into training and testing subsets several times, aid in evaluating the resilience and generalization capacity of the model. Additionally, model evaluation entails evaluating the effectiveness of various ML setups or algorithms to determine which method is best for jobs involving crop health monitoring [31]. This comparative study might involve testing different algorithms, hyperparameters, feature sets, or benchmarking against baseline models. When evaluating models in agricultural contexts, practical limitations like available data, processing power, and interpretability standards are also considered. Complicated deep learning models, for instance, may be quite accurate, but they would need a lot of computing power to train and infer, making it harder to use them in situations with limited resources. All things considered, a systematic and thorough review approach that considers both practical issues and performance indicators is necessary for an effective model assessment in crop health monitoring. Researchers and practitioners can refine and optimize ML models repeatedly depending on evaluation outcomes and provide dependable and strong solutions to monitor and control crop health efficiently.

1.3.3 BALANCING VARIANCE AND BIAS IN CROP HEALTH PREDICTION MODELS

Creating crop health prediction models that are accurate and dependable requires careful consideration of bias and variation. Within the domain of ML, bias denotes the inaccuracy resulting from excessively basic assumptions in the model, which causes underfitting [32], while variance refers to the model's susceptibility to variations in the training data, which causes overfitting. A model with a large bias in crop health prediction may be unable to accurately forecast crop health due to its inability to grasp the intricate interactions between environmental conditions and crop health indicators [33]. Conversely, a high variance model could be able to pick up on noise in the training set, which would hinder its ability to generalize to new data. Many strategies may be used to strike a balance between bias and variation. Employing ensemble techniques is one strategy, such as gradient boosting or random forests, which combine several weak learners to lower variance while preserving low bias. The purpose of ensemble [34]. Regularization encourages simpler models that generalize to new data more readily by including a penalty element in the model's loss function [34, 35]. By repeatedly dividing the dataset into training and validation sets, cross-validation may also be used to quantify bias and variance. Through an examination of the trade-off between variance and bias in various models or hyperparameters, practitioners may determine the best balance for their particular use case. In general, minimizing bias and variation in agricultural health forecasting models necessitates giving regularization methods, validation tactics, and model complexity considerable thought. Researchers and practitioners may create models that reliably forecast crop health indicators while preserving robustness and generalizability to new data by finding the ideal balance.

1.4 FUTURE TRENDS IN AI/ML TECHNIQUES FOR CROP HEALTH MONITORING

Future developments in AI/ML approaches for crop health monitoring have the potential to completely transform agriculture [36] by bringing in creative solutions that boost resilience, sustainability, and production. Precision agriculture will enter a new phase marked by integrating modern sensor technology, including drones, Internet of Things (IoT) devices, and hyperspectral photography. With the help of these sensors, farmers will be able to make well-informed decisions for optimal crop management by receiving high-resolution, real-time data on vital variables like soil moisture, nutrient levels, and insect infestations. Integrating multi-modal data from several sources, such as satellite images, weather predictions, and historical crop data, is another noteworthy [37] trend. In order to analyze this combined data and find complex patterns and connections that help provide more precise crop health predictions, AI/ML models will be essential. This comprehensive technique makes it possible to fully comprehend the dynamic elements affecting crop well-being. It is projected that explainable AI/ML models will play a bigger role in the agriculture industry [38]. To maintain decision-making process confidence, farmers and other stakeholders will need clear and understandable insights into AI-driven forecasts. With a focus on explainability, model forecasts will always have a clear explanation, empowering farmers to make confident and knowledgeable crop management decisions. An innovative development in crop health management is using AI/ML algorithms in autonomous agricultural systems, which automate crop health monitoring and management. Advanced algorithms will be utilized by these systems to examine sensor data, identify irregularities, and independently carry out remedial measures like focused pesticide application or precise watering. This independence maximizes crop health and facilitates agricultural processes in actual time. In the future, collaborative research and data sharing will be essential components of AI/ML in agriculture. More cooperation between data scientists, agronomists, and academics will result in common data repositories and cooperative platforms. With the goal of promoting information sharing, accelerating innovation, and advancing agricultural sustainability, this collaborative strategy hopes to support international efforts to address issues related to food security while reducing environmental effects [39].

1.4.1 ISSUES AND CHALLENGES IN AI/ML FOR AGRICULTURE

For AI/ML to be implemented effectively and widely, a number of concerns and obstacles related to its adoption in agriculture must be resolved. The availability and caliber of data provide a major obstacle [40]. Data compatibility, consistency, and completeness problems arise from the fact that agricultural data are frequently gathered from a variety of sources, including sensors, satellites, and historical records [41]. The accuracy and resilience of AI/ML models depend heavily on the quality and consistency of the data. The absence of infrastructure and technical preparedness in agricultural areas, especially in emerging nations, is another problem. Deploying AI/ML solutions in these places is hampered by limited access to modern technology, high-speed internet, and computer resources. It is crucial to close the digital gap and provide the infrastructure needed for data gathering, processing, and distribution [42]. For AI/ML to be implemented effectively and widely, a number of concerns and obstacles related to its adoption in agriculture must be resolved. The availability and caliber of data provide a major obstacle. Data compatibility [43], consistency, and completeness problems arise from the fact that agricultural data are frequently gathered from various sources, including sensors, satellites, and historical records.

The accuracy and resilience of AI/ML models depend heavily on the quality and consistency of the data [44]. The absence of infrastructure and technical preparedness in agricultural areas, especially in emerging nations, is another problem. Deploying AI/ML solutions in these places is hampered by limited access to modern technology [45], high-speed internet, and computer resources. It is crucial to close the digital gap and provide the infrastructure needed for data gathering, processing, and distribution. for fostering confidence and trust in agricultural systems powered [46] by AI. Significant hurdles also lie in the scalability and applicability of AI/ML solutions to various agricultural contexts and cropping systems. The vast variations in agricultural techniques across different areas, climates, and crops necessitate using adaptable AI/ML frameworks [47]. To maximize the impact of AI in agriculture, domain-specific and context-aware AI/ML models that adjust to local conditions and farmer preferences must be developed. Unlocking the full potential of AI/ML to transform agricultural sustainability and food security globally will depend on addressing these problems and obstacles.

1.4.2 EMERGING TRENDS IN AI/ML RESEARCH AND APPLICATIONS FOR CROP HEALTH MONITORING

New directions in AI/ML research and crop health monitoring applications have great potential to transform agriculture and improve food security. The growing use of drones and unmanned aerial vehicles (UAVs) [48] with sophisticated sensors for accurate and effective crop monitoring is one noteworthy development. With the use of these airborne platforms, farmers can now diagnose crop stress, disease outbreaks, and nutrient shortages with previously unheard-of accuracy [49]. These platforms also record high-resolution pictures and multispectral data. Creating AI-driven predictive analytics models for early pest and disease identification and forecasting is another new trend. These models use biological markers, historical data, and environmental factors to forecast the probability of disease outbreaks and suggest prompt actions to avoid crop losses [50]. Furthermore, the application of AI algorithms optimizes pest control tactics, such as the use of integrated pest management (IPM) techniques and biocontrol agents. Additionally, the use of explainable AI (XAI) approaches to improve the comprehensibility and transparency of AI/ML models in crop health monitoring is gaining traction [51]. With the use of XAI techniques, agronomists and farmers can comprehend the reasoning behind forecasts and recommendations by gaining insight into the decision-making process of intricate AI models. The adoption of AI-driven solutions in agricultural decision-making is facilitated by this openness, which also builds trust and confidence in these solutions. Furthermore, real-time crop health management and monitoring at the field level is made possible by the combination of AI/ML with edge computing and IoT sensors [52]. IoT sensors incorporated into plants, soil, and farming machinery continuously gather data on characteristics related to crop health, moisture content, and the surrounding environment. Real-time data analysis using AI algorithms helps farmers make data-driven decisions and maximize resource use for higher agricultural yields and financial success.

1.5 CONCLUSION

In conclusion, crop health monitoring and management might be completely transformed by AI/ML approaches, which would provide farmers with never-before-seen tools and insights to maximize agricultural operations. The discipline of AI/ML offers a wide range of tools for evaluating massive volumes of agricultural data and deriving useful insights, from supervised and unsupervised learning techniques to RL and deep learning algorithms. In this chapter, we looked at a number of AI/ML applications for crop health monitoring, such as yield prediction, pest control, and disease detection. The significance of high-quality data and data pre-treatment were deliberated upon alongside pragmatic aspects like feature engineering and model evaluation. We also looked at new developments in AI/ML research, such as using UAVs in conjunction with predictive analytics, explainable AI, and Internet of Things-driven real-time monitoring and management systems of the health of the crop. Notwithstanding AI/ML's enormous promise in agriculture, there are still issues and restrictions that must be resolved. Careful management of issues like algorithm bias, data privacy, and the digital divide is necessary to provide fair access to AI-driven solutions and avoid unforeseen repercussions. Furthermore, in order to create reliable and flexible AI/ML models, multidisciplinary cooperation and continuous study are necessary due to the complex and dynamic character of agricultural systems. Crop health monitoring has a bright future ahead of it because of the rapid improvements in AI and ML technologies. Farmers can increase crop yields, optimize resource use, and improve sustainability across agricultural landscapes by utilizing AI and ML. AI/ML has the ability to bring about revolutionary change in agriculture and tackle the changing issues of environmental sustainability and food security.

REFERENCES

[1]   FAO: Agricultural Production Statistics 2000–2021. Available at: fao.org/3/cc3751en/cc3751en.pdf [Accessed 20 June 2023].
[2]   Food and Agriculture Organization of the United Nations: Climate Change and Food Security: Risks and Responses. Available at: fao.org/3/i5188e/I5188E.pdf [Accessed 14 June 2023].
[3]   K. Suzuki (Ed.), Artificial Neural Networks – Industrial and Control Engineering Applications, In-Tech, Croatia, 2011.
[4]   S. Kujawa, G. Niedbała (Eds.), *Artificial Neural Networks in Agriculture*, MDPI, 2021. doi.org/10.3390/books978-3-0365-1579-3. Available at: www.mdpi.com/journal/agriculture/special issues/Artificial Neural Networks Agriculture
[5]   E. Funes, Y. Allouche, G. Beltrán, A. Jiménez, A Review: Artificial Neural Networks as Tool for Control Food Industry Process, *Journal of Sensor Technology*, 5, 2015, pp. 28–43. doi.org/10.4236/jst.2015.51004
[6]   S. Francik, Z. Slipek, J. Fraczek, A. Knapczyk, Present Trends in Research on Application of Artificial Neural Networks in Agricultural Engineering, *Agricultural Engineering*, 20 (4), 2017, pp. 15–25. doi.org/10.1515/agriceng-2016-0060
[7]   R. Lopez-Lozano, B. Baruth, An Evaluation Framework to Build a Cost-Efficient Crop Monitoring System. Experiences from the Extension of the European Crop Monitoring System, *Agricultural Systems*, 168, 2019, pp. 231–246. doi.org/10.1016/j.agsy.2018.04.002

[8] L. Rutkowski, Generalized Regression Neural Networks in Time-Varying Environment, *IEEE Transactions on Neural Networks,* 15 (3), 2004, pp. 576–596. doi.org/10.1109/TNN.2004.826127

[9] L. Rutkowski, Adaptive Probabilistic Neural Networks for Pattern Classification in Time-Varying Environment, *IEEE Transactions on Neural Networks,* 15 (4), 2004, pp. 811–827. doi.org/10.1109/TNN.2004.828757

[10] FAOSTAT: Crop and Livestock Products. Available at: fao.org/faostat/en/#data/QCL [Accessed 18 June 2023].

[11] M. Pérez-Pons, J. Parra-Dominguez, S. Omatu, E. Herrera-Viedma, J. Corchado, Machine Learning and Traditional Econometric Models: A Systematic Mapping Study, *Journal of Artificial Intelligence and Soft Computing Research*, 12 (2), 2022, pp. 79–100. doi.org/10.2478/jaiscr-2022-0006

[12] M.F. Mushtaq, U. Akram, M. Aamir, H. Ali, M. Zulqarnain, Neural Network Techniques for Time Series Prediction: A Review, *JOIV: International Journal on Informatics Visualization,* 3 (3), 2019, pp. 314–320. dx.doi.org/10.30630/joiv.3.3.281

[13] J. Lv, M. Pawlak, Bandwidth Selection for Kernel Generalized Regression Neural Networks in Identification of Hammerstein Systems, *Journal of Artificial Intelligence and Soft Computing Research*, 11 (3), 2021, pp. 181–194. https://doi.org/10.2478/jaiscr-2021-0011

[14] T.J. Ross, J.M. Booker, W.J. Parkinson (Eds.), Fuzzy Logic and Probability Applications: Bridging the Cap, ASA-SIAM Series on Statistics and Applied Probability, SIAM, Philadelphia, ASA, Alexandria, VA, 2002. URL: pzs.dstu.dp.ua/logic/bibl/prob.pdf

[15] X. Li, M.J. Er, B.S. Lim, J.H. Zhou, O.P. Gan, L. Rutkowski, Fuzzy Regression Modeling for Tool Performance Prediction and Degradation Detection, *International Journal of Neural Systems*, 20 (5), 2010, pp. 405–419. doi.org/10.1142/S0129065710002498

[16] I. Laktionov, O. Vovna, M. Kabanets, Computer-Oriented Method of Adaptive Monitoring and Control of Temperature and Humidity Mode of Greenhouse Production, *Baltic Journal of Modern Computing*, 11 (1), 2023, pp. 202–225. doi.org/10.22364/bjmc.2023

[17] I. Laktionov, O. Vovna, M. Kabanets, Information Technology for Comprehensive Monitoring and Control of the Microclimate in Industrial Greenhouses Based on Fuzzy Logic, *Journal of Artificial Intelligence and Soft Computing Research*, 13 (1), 2023, pp. 19–35. doi.org/10.2478/jaiscr-2023-0002

[18] L. Rutkowski, Flexible Neuro-Fuzzy Systems, Springer New York, NY, 2004. doi.org/10.1007/b115533

[19] D. Rutkowska, Neuro-Fuzzy Architectures and Hybrid Learning, Physica Heidelberg, 2002. doi.org/10.1007/978-3-7908-1802-4

[20] P. Dziwiński, A. Przybył, P. Trippner, J. Paszkowski, Y. Hayashi, Hardware Implementation of a Takagi-Sugeno Neuro-Fuzzy System Optimized by a Population Algorithm, *Journal of Artificial Intelligence and Soft Computing Research*, 11 (3), 2021, pp. 243–266. doi.org/10.2478/jaiscr-2021-0015

[21] A. Niewiadomski, M. Kacprowicz, Type-2 Fuzzy Logic Systems in Applications: Managing Data in Selective Catalytic Reduction for Air Pollution Prevention, *Journal of Artificial Intelligence and Soft Computing Research*, 11 (2), 2021, pp. 85–97. doi.org/10.2478/jaiscr-2021-0006

[22] D. Karaboga, E. Kaya, Adaptive Network Based Fuzzy Inference System (ANFIS) Training Approaches: A Comprehensive Survey, *Artificial Intelligence Review*, 52, 2019, pp. 2263–2293. doi.org/10.1007/s10462-017-9610-2

[23] Y.K. Al-Douri, H. Hamodi, J. Lundberg, Time Series Forecasting Using a Two-Level Multi-Objective Genetic Algorithm: A Case Study of Maintenance Cost Data for Tunnel Fans, *Algorithms*, 11 (123), 2018, pp. 1–19. doi.org/10.3390/a11080123

[24] I. Aouadni, A. Rebai, Decision Support System Based on Genetic Algorithm and Multi-Criteria Satisfaction Analysis (MUSA) Method for Measuring Job Satisfaction, *Annals of Operations Research*, 256, 2017, pp. 3–20. doi.org/10.1007/s10479-016-2154-z

[25] H. Han, J. Siebert, TinyML: A Systematic Review and Synthesis of Existing Research, In: 2022 International Conference on Artificial Intelligence in Information and Communication, Jeju Island, Republic of Korea, 2022, pp. 269–274. doi.org/10.1109/ICAIIC54071.2022.9722636

[26] P.P. Ray, A Review on TinyML: State-of-the-Art and Prospects, *Journal of King Saud University – Computer and Information Sciences*, 34 (4), 2022, pp. 1595–1623. doi.org/10.1016/j.jksuci.2021.11.019

[27] N. Schizas, A. Karras, C. Karras, S. Sioutas, TinyML for Ultra-Low Power AI and Large Scale IoT Deployments: A Systematic Review, *Future Internet*, 14 (363), 2022, pp. 1–45. doi.org/10.3390/fi14120363

[28] G. Schwalbe, B. Finzel, A Comprehensive Taxonomy for Explainable Artificial Intelligence: A Systematic Survey of Surveys on Methods and Concepts, *Data Mining and Knowledge Discovery*, 2023, pp. 1–59. doi.org/10.1007/s10618-022-00867-8

[29] W. Saeed, C. Omlin, Explainable AI (XAI): A Systematic Meta-Survey of Current Challenges and Future Opportunities, *Knowledge-Based Systems*, 263, 2023, pp. 1–24. doi.org/10.1016/j.knosys.2023.110273

[30] G. Vilone, L. Longo, Notions of Explainability and Evaluation Approaches for Explainable Artificial Intelligence, *Information Fusion*, 76, 2021, pp. 89–106. doi.org/10.1016/j.inffus.2021.05.009

[31] P. Mishra, Explainable AI Recipes: Implement Solutions to Model Explainability and Interpretability with Python, Springer, 2023. ISBN-13 (pbk): 978-1-4842-9028-6, doi.org/10.1007/978-1-4842-9029-3

[32] M. Ryo, Explainable Artificial Intelligence and Interpretable Machine Learning for Agricultural Data Analysis, *Artificial Intelligence in Agriculture,* 6, 2022, pp. 257–265. doi.org/10.1016/j.aiia.2022.11.003

[33] S. Mohr, R. Kühl, Acceptance of Artificial Intelligence in German Agriculture: An Application of the Technology Acceptance Model and the Theory of Planned Behavior, *Precision Agriculture*, 22, 2021, pp. 1816–1844. doi.org/10.1007/s11119-021-09814-x

[34] N.L. Tsakiridis, T. Diamantopoulos, A.L. Symeonidis, J.B. Theocharis, A. Iossifides, P. Chatzimisios, G. Pratos, D. Kouvas, Versatile Internet of Things for Agriculture: An eXplainable AI Approach, *Artificial Intelligence Applications and Innovations*, 584, 2020, pp. 180–191. doi.org/10.1007/978-3-030-49186-4

[35] I. Ahmed, G. Jeon, F. Piccialli, From Artificial Intelligence to Explainable Artificial Intelligence in Industry 4.0: A Survey on What, How, and Where, *IEEE Transactions on Industrial Informatics*, 18 (8), 2022, pp. 5031–5042. doi.org/10.1109/TII.2022.3146552

[36] S. Rakesh, M. Indiramma, Explainable AI for Crop Disease Detection, In: 2022 4th International Conference on Advances in Computing, Communication Control and Networking, Greater Noida, India, 2022, pp. 1601–1608. doi.org/10.1109/ICAC3N56670.2022.10074303

[37] H. Oubehar, A. Selmani, A. Ed-Dahhak, A. Lachhab, M. El Hassane Archidi, B. Bouchikhi, ANFIS-Based Climate Controller for Computerized Greenhouse System, *Advances in Science, Technology and Engineering Systems Journal,* 5 (1), 2020, pp. 8–12. doi.org/10.25046/aj050102

[38] V. Vincentdo, N. Surantha, Nutrient Film Technique-Based Hydroponic Monitoring and Controlling System Using ANFIS, *Electronics,* 12 (1446), 2023, pp. 1–26. doi.org/10.3390/electronics12061446

[39] D. Bozanic, D. Tesic, D. Marinkovic, A. Milic, Modeling of Neuro-Fuzzy System as a Support in Decision-Making Processes, *Reports in Mechanical Engineering*, 2 (1), 2021, pp. 222–234. doi.org/10.31181/rme2001021222b

[40] N. Ziasabounchi, I.N. Askerzade, ANFIS Based Classification Model for Heart Disease Prediction, *International Journal of Engineering & Computer Science IJECS-IJENS*, 14 (02), 2014, pp. 7–12, URL: ijens.org/IJECSVol14Issue02.html

[41] S. Bellahirich, D. Mezghani, A. Mami, Design and Implementation of an Intelligent ANFIS Controller on a Raspberry Pi Nano-Computer for Photovoltaic Pumping Intended for Drip Irrigation, *Energies*, 14 (5217), 2021, pp. 1–19. doi.org/10.3390/en14175217

[42] A.S. Keceli, A. Kaya, C. Catal, B. Tekinerdogan, Deep Learning-Based Multi-Task Prediction System for Plant Disease and Species Detection, *Ecological Informatics*, 69, 2022, pp. 1–14. doi.org/10.1016/j.ecoinf.2022.101679

[43] S.H. Lee, H. Goëau, P. Bonnet, A. Joly, Conditional Multi-Task Learning for Plant Disease Identification, In: 2020 25th International Conference on Pattern Recognition, Milan, Italy, 2021, pp. 3320–3327. doi.org/10.1109/ICPR48806.2021.9412643

[44] M.S. Hema, N. Sharma, Y. Sowjanya, Ch. Santoshini, R. Sri Durga, V. Akhila, Plant Disease Prediction Using Convolutional Neural Network, *EMITTER International Journal of Engineering Technology*, 9 (2), 2021, pp. 283–293. doi.org/10.24003/emitter.v9i2.640

[45] T.V. Reddy, K.S. Rekha, Plant Disease Detection Using Advanced Convolutional Neural Networks with Region of Interest Awareness, *Stem Cell Research International*, 6 (2), 2022, pp. 121–131. doi.org/10.47363/JIRR/2022(2)117

[46] T. Domingues, T. Brand̄ao, J.C. Ferreira, Machine Learning for Detection and Prediction of Crop Diseases and Pests: A Comprehensive Survey, *Agriculture*, 12 (1350), 2022, pp. 1–23. doi.org/10.3390/agriculture12091350

[47] H. Bangui, S. Rakrak, S. Raghay, B. Buhnova, Moving to the Edge-Cloud-of-Things: Recent Advances and Future Research Directions, *Electronics*, 7 (11:309), 2018, pp. 1–31. doi.org/10.3390/electronics7110309

[48] I.S. Laktionov, O.V. Vovna, M.M. Kabanets, H.O. Sheina, I.A. Getman, Information Model of the Computer-Integrated Technology for Wireless Monitoring of the State of Microclimate of Industrial Agricultural Greenhouses, *Instrumentation Mesure Metrologie*, 20 (6), 2021, pp. 289–300. doi.org/10.18280/i2m.200601

[49] I. Laktionov, L. Rutkowski, O. Vovna, A. Byrski, M. Kabanets, A Novel Approach to Intelligent Monitoring of Gas Composition and Light Mode of Greenhouse Crop Growing Zone on the Basis of Fuzzy Modelling and Human-in-the-Loop Techniques, *Engineering Applications of AI*, 126 (B), 2023, pp. 1–21, doi.org/10.1016/j.engappai.2023.106938

[50] Math Works: Nonlinear System Identification. Available at: mathworks.com/help/fuzzy/nonlinear-system-identification

[51] University of Illinois: Report on Plant Diseases. Common Leaf Blights and Spots of Corn. Available at: ipm.illinois.edu/diseases/rpds/2022

[52] Crop Science Australia: Fusarium Head Blight. Available at: crop.bayer.com.au/pests/diseases/fusarium-head-blight

2 Enhancing Crop Productivity by Suitable Crop Prediction Using Cutting-Edge Technologies

G. Mamatha and Jyothi S. Nayak

2.1 INTRODUCTION

Agriculture provides a significant resource for people living in rural areas. World economies rely on agricultural products. In 2011, the Horticultural Statistics of India reported that out of 1300 million people in India, almost 61.5% were involved in some form of agriculture or horticulture. There are 159.6 million of these households. 20.5% of India's gross domestic product is allocated to the agricultural sector. A large portion of India's population works in the agriculture sector. About eighty percent of people living in rural areas rely on agriculture workers.

With every passing day, the need for food supplies is rising sharply. The United Nations Food and Agriculture Organization has projected that India's population will reach around 12 billion by the year 2050, necessitating a 60% rise in food production. [1]. Among all industrial endeavors, agriculture is among the most fundamental and long-standing. We are all aware that agriculture is the foundation of India. Numerous Indian farmers, however, continue to rely on traditional methods. It affects the financial sector.

In the past, farmers had to rely on their rudimentary knowledge when choosing which crops to utilize. Farmers usually prefer to grow whichever crop is popular in the vicinity. Soil nutrients are depleted due to insufficient crop rotation and an absence of agricultural knowledge. The agricultural sector must become more tech-savvy and digital. The importance of soil nutrients, groundwater level, and pesticide type in determining crop quality [2] requires careful consideration. Land and water are the two most important factors influencing agricultural production. Agricultural resources are depicted in Figure 2.1.

Agriculture faces numerous critical challenges. Agricultural practices are complex and reliant on a wide variety of inputs and outputs, including but not limited to water,

DOI: 10.1201/9781003484608-2

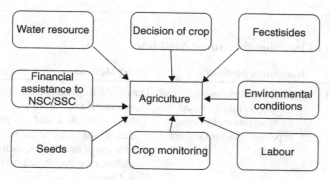

FIGURE 2.1   Resources of agriculture.

labor, crop selection, weather (including rainfall, temperature, and humidity), crop monitoring, pest control, and applying the appropriate amount of pesticides [3].

Blockchain technology, cloud computing, and deep learning are emerging trends in the computer field, with applications in various domains. Deep learning algorithms are making machine learning more powerful and accurate. Auto-harvesting robots, machine learning, and deep learning techniques are helping farmers reduce losses in the harvesting stage. However, challenges in implementing machine learning algorithms in agriculture include data requirements and selecting suitable algorithms from a wide list of available algorithms. Despite these challenges, advancements in machine learning and deep learning models enhance agricultural efficiency [4].

2.2   TRADITIONAL METHODS VS SMART FARMING

Monitoring crops is crucial. In traditional farming, farmers spend more time in their fields and visit them daily. However, crops are impacted by environmental conditions and other factors. Farmers suffer more losses due to sudden weather changes and other causes. Traditional farming practices, which are widely used, result in low agricultural yields. Table 2.1 discusses key concepts in terms of traditional and smart farming methods.

Traditional farming operations rely more on traditional methods and physical labor than modern technology or data-driven approaches. The excessive use of pesticides in traditional methods affects soil and human health. Most farmers use pesticides without realizing the harm they cause to the soil, ecosystem, and human health. Traditional methods involve applying pesticides directly to the soil without considering the soil's composition or the impact of diseases, leading to the use of a greater quantity of pesticides.

Conversely, "smart farming" refers to a farming method that extensively uses data and cutting-edge technology. The goal is to maximize output and resource efficiency while lowering environmental impact through data-driven decision-making, sustainable practices, automation, and accuracy.

One example is the application of sensors in the implementation of smart pesticides. Depending on the pesticide application method, nitrogen, phosphorus, and potassium

TABLE 2.1
Key Aspects of Traditional vs Smart Methods

Key Aspects	Traditional Method	Smart Method
Based on data approach	Decision-making in traditional farming is mostly based on observation, Personal experience and traditional methods.	Data collecting, analysis, and interpretation are crucial to smart farming. Sensors, drones, satellites, and computer vision are utilized in this to gather data on a wide range of subjects, including soil quality, crop health, seasonal changes, and more.
Precision and accuracy	The applied techniques and estimations used in traditional farming may result in inefficiencies and a waste of resources.	Precision and accuracy are key components of smart farming. It precisely calculates the right amount of inputs, which include water, nitrogen fertilizer, and pesticides, required for appropriate crop growth by implementing real-time data and AI algorithms.
Robots and automated systems	Several tasks in traditional farming are performed manually by farmers or laborers, which can be mentally as well as physically challenging.	Robotics and automation are used in smart farming to complete labor-intensive Chores. Robots and autonomous devices can perform tasks like planting, weeding, and Harvesting with a high degree of precision and effectiveness.
Monitoring and control of crop	Observation and physical presence are usually used in traditional farming to monitor and control operations.	Smart farming allows farmers to remotely track and coordinate various elements associated with their operations. Using IoT devices, sensors, and automated systems, farmers can monitor temperature, humidity, and other soil features, and make real-time adjustments even when not physically present on the farm.
Decision support systems	Traditional farming relies more on the knowledge and expertise of the farmers, which may be Constrained or biased.	Decision support systems in smart farming use AI and ML algorithms to analyze data from various sources, providing insights that support decision-making for multiple agricultural operations.
Efficiency of resources and sustainability	Resource effectiveness and environmental sustainability may not be Prioritized in traditional farming practices.	Smart farming aims to promote sustainable agricultural practices and maximize resource utilization. By employing data-driven initiatives, farmers can reduce their usage of water, fertilizer, and pesticides, positively impacting the environment.

(NPK) sensors provide information regarding NPK when leaves are dry or chlorotic. Insect-infested areas are sprayed with insecticides, reducing cost and harm compared to traditional methods, which use more pesticides and are more expensive.

To describe current trends involving the use of AI in smart agriculture, the latest technologies, the generation of data, the processing of data, and the path from traditional to smart agriculture [5].

Traditional farming involves farmers visiting their fields to understand crop conditions. The agricultural era 2.0 began with the use of machinery and chemicals, improving efficiency and productivity. Smart farming automates processes like harvesting and crop yields, enabling data acquisition, processing, monitoring, decision-making, and management. Financial resources can help farmers in times of unexpected calamities. Education and digital literacy are crucial for farmers to effectively use smart farming technologies and understand their limitations [6].

2.3 RELATED WORK

2.3.1 AGRICULTURE IN IoT

Farms rely on water as their principal resource. Utilizing sensors can maintain efficient water management. The Internet of Things (IoT) aids in the early detection and mitigation of diseases and damage by continuously monitoring crops. To take early measures, it is beneficial to continuously monitor crops. This enables farmers to take early measures, increasing output while decreasing the time and effort spent on human labor and associated costs. Sensors in agricultural soil management can determine the soil's pH and moisture level, enabling farmers to sow seeds at the optimal depth.

The study uses an IoT-based monitoring system to monitor the impact of physical conditions on plant growth. Data is collected from various sensors, including humidity, temperature, soil temperature, moisture, and light intensity. Logistic Regression, Gradient Boosting Classifier, and Linear Support Vector Classifier algorithms are used to analyze these parameters. The system architecture includes image processing and an open CV library for data collection. The data is then sent to the cloud for storage and monitoring. The model can detect plant growth rates by analyzing the environmental conditions [7].

2.3.2 IMPORTANCE OF MACHINE LEARNING IN AGRICULTURE

The role of machine learning (ML) in the agricultural sector is growing, offering many benefits that might revolutionize farming. Machine learning has proven essential in several major areas of agriculture:

i) **Yield Prediction and Optimization:** By considering parameters such as weather patterns, soil quality, and crop health, ML models can offer insight into optimal planting intervals, watering needs, and crop rotation strategies. This helps farmers improve yields and make cost-effective operational plans.

ii) **Crop and Soil Monitoring:** Machine learning-driven systems can track and analyze soil conditions, crop health, and growth trends. ML algorithms can detect nutritional deficiencies, insect infestations, and early warning signs of diseases through computer vision and machine learning. This allows farmers to avoid crop losses, treat specific areas, and act quickly.

The primary objective of this paper is to shed light on the application of machine learning in agriculture by conducting a thorough review of recent scholarly literature, focusing on keywords related to crop management, water management, soil management, and livestock management within the years 2018–2020 [8].

A soil nutrient system and an insecticide recommender system were used in a study where data sets of five different kinds of pests commonly attacking crops were gathered. A convolutional neural network (CNN) and machine vision were employed to determine which pests to spray on crops. The NPK sensor records the map's readings and sends out a warning message to suggest pesticides if the nutritional values are low. Data was collected from official websites and conversations with farmers [9][10].

Traditional methods of applying fertilizers involve farmers combining several types of fertilizers based on their knowledge and financial situation rather than the soil's specific needs. This impacts soil and crop development. A fertilizer suggestion system was thus necessary for them. A mechanism for recommending fertilizer was developed and applied in the article presented in [11]. To address this, a fertilizer recommendation system was developed using a bi-LSTM design. This algorithm suggests fertilizer combinations such as DAP, MOP, and UREA, determining the optimal ratio to minimize resource consumption based on the details recorded by farmers.

A novel system design supporting an IoT-enabled irrigation drip system, operable via an Android or web app, was discussed in another study. The Node MCU communicates information to the network cloud system using wireless communications. Data can be accessed remotely through web apps after a successful IoT cloud exchange. The automated pumping process supplies a precise quantity of water, allowing plants to thrive and enabling farmers to monitor their condition from any location using an Android smartphone. This offers a practical and economical means of accessing water supplies, mitigating concerns about water scarcity [12].

Intelligent watering systems were also discussed, where semi-supervised learning algorithms using the NODE RED platform and other sensor-based techniques ensure accurate decision-making [13].

Crop selection is crucial for maximizing yield, but a lack of knowledge about soil fertility and crop selection leads to low production. A model based on crop selection, management, and maturity is proposed, with a pH crop prediction method and real-time data analysis. A new data mining approach combines climate and crop production data for yield prediction [14].

Another study focused on determining optimal conditions for marigold plant growth using various sensors to measure soil moisture, environmental temperature, light intensity, humidity, and soil temperature. These values are stored in the cloud, and the plant's dimensions are also recorded [15].

Research on apple harvest management addressed the codling moth pest problem. The system applies treatments for the codling moth only when a threat to crops is detected, maximizing the use of insecticides while minimizing environmental impact [16].

A paper proposed a drone application technique that uses the IoT framework to classify rice diseases based on real-time data, data gathering, and analytic methods. The system uses the Global Positioning System (GPS) to pinpoint the location of diseased rice plants in fields for precise treatment [17].

The paper introduces a smart agriculture watering system using IoT sensors to monitor soil moisture levels, optimize irrigation, and predict evapotranspiration, demonstrating its potential to enhance crop yields [18].

All the above studies are based on the Internet of Things to improve the soil nature and recommendation system for crops. In this research paper, we present a novel crop recommendation approach based on several parameters related to the crop and soil collected over IoT sensors deployed in the field, using suitable machine learning algorithms to maximize crop yield.

2.4 PROPOSED METHODOLOGY

The proposed crop recommendation method uses machine algorithms – Random Forest and Support Vector Machine – applied to a data set collected using an IoT setup.

Integrating machine learning in farming can significantly enhance crop yields, irrigation, and other agricultural operations by predicting or classifying outcomes based on historical data. Users can utilize trained machine learning models to make real-time predictions or suggestions. Farmers and other stakeholders in the agricultural sector can access these forecasts and recommendations through various services and mobile applications. The proposed method integrates data collection and transmission through IoT devices with machine learning algorithms to facilitate this process. The process flow of the proposed system is depicted in Figure 2.2.

2.4.1 DATA SET

The process of preparing and cleaning the CSV-formatted fertilizer datasets for training in a data frame has been completed. The training dataset comprises 80% of the data, while the test dataset accounts for the remaining 20% [19]. Each dataset contains 100 tuples encompassing all the specified soil characteristics. Subsequently, machine learning models are employed on these datasets to recommend suitable fertilizers based on the soil features [20]. This approach aims to leverage data-driven insights for precision in fertilizer recommendations, enhancing agricultural practices.

The dataset used to implement the proposed crop recommendation system was collected from the agriculture office of Andhra Pradesh. A sample test set is shown in Table 2.2. The dataset includes several features like pH, EC, sulfur, nitrogen, phosphorus, potassium, zinc, and boron. Figure 2.3 displays a correlation matrix as a heatmap, providing a color-coded depiction of the variables for intuitive understanding.

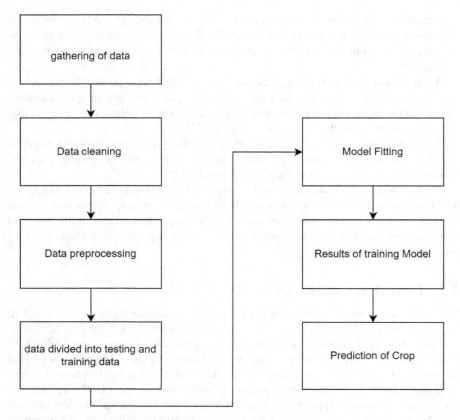

FIGURE 2.2   Process flow of the proposed system.

TABLE 2.2
Sample Data Set

S.No	Parameter	Test Value	Unit	Rating
1	pH	8.3		alkaline
2	EC	0.35	dS/m	Normal
3	Organic Carbon (OC)	0.17	%	Low
4	Nitrogen	18	Kg/ha	Low
5	Phosphorous	23	Kg/ha	Medium
6	Potassium	154	Kg/ha	Sufficient
7	Sulfur	24	ppm	Deficient
8	Zinc	1.56	ppm	Low
9	Boron	0,63	ppm	High

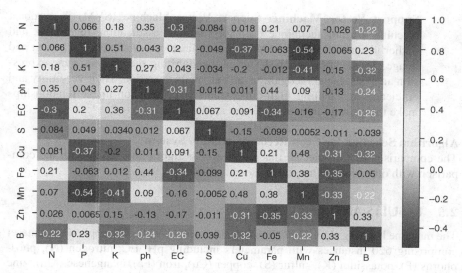

FIGURE 2.3 Correlation matrix between different variables.

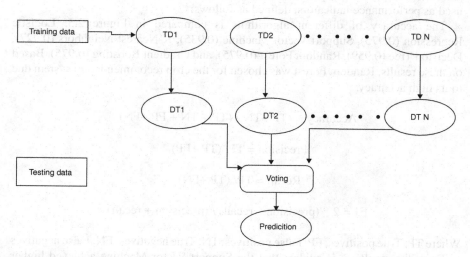

FIGURE 2.4 The working of the Random Forest algorithm.

2.4.2 MACHINE LEARNING ALGORITHMS USED

Two machine learning algorithms, Support Vector Machine and Random Forest, are applied on the above data set.

Random Forest: A Random Forest improves prediction accuracy by averaging the results of a large number of decision trees, each trained on different subsets of the dataset. It uses the combined predictions of all trees to answer the problem, tallying up the votes of the most forecasts. Figure 2.4 explains the working of the Random Forest algorithm [21].

i) **Support Vector Machine:** Support Vector Machine (SVM) is a robust machine learning technique used for various classification, regression, and outlier identification tasks. SVMs are applicable in contexts such as text and image classification, spam detection, handwriting recognition, gene expression analysis, facial recognition, and anomaly detection. Their versatility and efficiency in handling high-dimensional data and non-linear relationships make them useful in many different applications.

Algorithm Selection for Crop Recommendation System

The comparison of the selected machine learning algorithms, along with the comparison with other algorithms, can be seen in Table 2.3.

2.5 RESULTS AND ANALYSIS

The machine learning algorithms, Random Forest and SVM, were applied to a dataset comprising 620 instances with parameters including pH, EC, nitrogen (N), phosphorus (P), potassium (K), sulfur (S), copper (Cu), iron (Fe), manganese (Mn), zinc (Zn), and boron (B). The performance metrics used for evaluating these methods in predicting crop yield are presented in Table 2.4. Precision, recall, and F1 score are used as performance indicators, defined as follows:

The accuracy of different algorithms is compared in Figure 2.5: Logistic Regression (0.975), Support Vector Machine (0.935), K-Nearest Neighbors (0.927), Decision Tree (0.959), Random Forest (0.975), and Gradient Boosting (0.975). Based on these results, Random Forest was chosen for the crop recommendation system due to its high accuracy.

$$\text{Accuracy} = (TP + TN) / (TP + TN + FP + FN)$$

$$\text{Precision} = TP / (TP + FP).$$

$$\text{Recall} = TP / (TP + FN).$$

$$\text{F1} = 2 * (\text{precision} * \text{recall}) / (\text{precision} + \text{recall})$$

Where TP: True positives, FP: False positives; TN: True negatives; FN: False negatives

From the results, it is evident that the Support Vector Machine achieved higher precision (94.0%) and recall (97.9%) compared to the Random Forest, which had a precision of 89.2% and recall of 94.7%. However, the F1 score, which balances precision and recall, was slightly higher for SVM (95.9%) than for Random Forest (91.9%). The same is depicted in Figure 2.6 with a graph of Performance Indication – Precision, Recall and F1Score of applied machine learning algorithms.

The recommendation system utilizes levels of nitrogen (N), phosphorus (P), potassium (K), pH, EC, sulfur (S), copper (Cu), iron (Fe), manganese (Mn), zinc (Zn), and boron (B) to suggest suitable crops based on the trained data. For instance, given the following values: pH 6, EC 1.45 dS/m; and sulfur 0.28, copper 14.47, iron 179.63,

TABLE 2.3
Comparison with Other Machine Learning Algorithms: Strengths, Limitations, Performance, and Reason for Selection

Algorithm	Strengths	Limitations	Performance	Reason for Selection
Random Forest	• Robustness • Accuracy • Versatility • Feature importance • Performance	• Varied techniques in handling multiclass classification • Complexity in tuning and computational resources requirements	Accuracy: 97.5%	High accuracy, robustness, and feature importance determination, suitable for agricultural decision-making
Support Vector Machine	• Effectiveness in High Dimensional Spaces • Margin Maximization • Flexibility • Performance	• Slightly lower accuracy compared to Random Forest • Complexity in tuning and computational resources requirements	Accuracy: 93.5%	Handling high-dimensional spaces and non-linear relationships in agricultural datasets
Logistic Regression	• Simplicity and ease of implementation • Interpretability • Performance in binary classification problems	• Less effective for complex datasets with non-linear relationships	Accuracy: 97.5%	Lacks ability for multiclass classification, limiting applicability to agricultural data complexities
K-Nearest Neighbors (KNN)	• Simplicity and ease of implementation • Flexibility	• Computationally expensive for large datasets • Lower accuracy compared to Random Forest and SVM	Accuracy: 92.7%	Inefficiency in handling large, high-dimensional datasets
Decision Tree	• Easy interpretation and visualization • Handling of numerical and categorical data	• Prone to overfitting, especially with small datasets • Outperformed by Random Forest	Accuracy: 95.9%	Overfitting tendency limits reliability for broader agricultural recommendations
Gradient Boosting	• Stage-wise model building • Bias reduction • Performance	• Slower training and complex tuning compared to Random Forest • Higher computational complexity and tuning requirements	Accuracy: 97.5%	Powerful but less practical for real-time agricultural applications due to complexity and resource demands

TABLE 2.4
Performance Indication

S.No	Algorithm Name	Precision	Recall	FI Score
1	Support Vector Machine	0.940	0.979	0.959
2	Random Forest algorithm	0.892	0.947	0.919
3	K-Nearest Neighbors	0.92	0.92	0.93
4	Decision Tree	0.88	0.85	0.85
5	Gradient Boosting	0.90	0.91	0.91
6	Logistic Regression	0.85	0.84	0.84

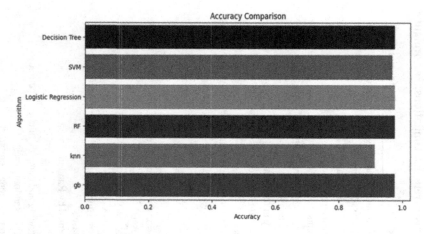

FIGURE 2.5 Accuracy comparison.

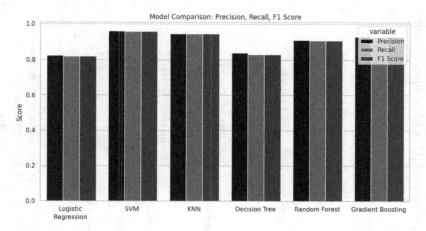

FIGURE 2.6 Performance Indication – Precision, Recall and F1Score of applied machine learning algorithms.

manganese 85.79, zinc 47.76, phosphorus 85, potassium 191, and boron 65.83 ppm, the system recommends mango as a suitable crop.

2.6 CONCLUSION

This article summarizes the findings from a machine learning-based research project aimed at developing crop recommendation models for predicting crop yields using SVM and Random Forest algorithms. Among these, the Random Forest model emerged as the superior choice. The approach demonstrated flexibility in its applicability to different datasets, regions, and countries.

The study's outcomes are poised to positively impact the agricultural sector in several ways. By assisting farmers in making informed crop choices, the models contribute to better agricultural practices. They also support the formulation of policies that benefit the sector, foster innovation in agricultural products and services, and contribute to price stability in agricultural markets. The scalability, accuracy, and user-friendly nature of the developed models make them valuable assets for farmers and stakeholders in the agricultural industry.

Future scope: Research on crop recommendation models using SVM and Random Forest algorithms shows potential for further advancement. The integration of IoT and AI can enhance the accuracy and responsiveness of crop recommendations. Expanding the scope of research beyond current limitations would enrich the dataset used for training models and ensure recommendations are tailored to specific agronomic challenges and environmental conditions. Incorporating climate data into predictive models could develop resilient agricultural practices, contributing to food security and economic stability on a global scale. These advancements aim to optimize agricultural productivity and foster sustainable practices.

REFERENCES

1. Calicioglu, Ozgul, et al. "The future challenges of food and agriculture: An integrated analysis of trends and solutions." *Sustainability* 11.1 (2019): 222.
2. Singh, Richa, Sarthak Srivastava, and Rajan Mishra. "AI and IoT-based monitoring system for increasing the yield in crop production." *2020 International Conference on Electrical and Electronics Engineering (ICE3)*. IEEE, 2020.
3. Brunelli, Davide, et al. "Energy neutral machine learning based IoT device for pest detection in precision agriculture." *IEEE Internet of Things Magazine* 2.4 (2019): 10–13.
4. Meshram, Vishal, et al. "Machine learning in agriculture domain: A state-of-art survey." *Artificial Intelligence in the Life Sciences* 1 (2021): 100010.
5. Javaid, Mohd, et al. "Understanding the potential applications of Artificial Intelligence in Agriculture Sector." *Advanced Agrochem* 2.1 (2023): 15–30.
6. Dhanaraju, Muthumanickam, et al. "Smart farming: Internet of Things (IoT)-based sustainable agriculture." *Agriculture* 12.10 (2022): 1745.
7. Singh, Richa, Sarthak Srivastava, and Rajan Mishra. "AI and IoT-based monitoring system for increasing the yield in crop production." *2020 International Conference on Electrical and Electronics Engineering (ICE3)*. IEEE, 2020.

8.   Benos, Lefteris, et al. "Machine learning in agriculture: A comprehensive updated review." *Sensors* 21.11 (2021): 3758.

9.   Khan, Arfat Ahmad, et al. "Internet of things (IoT)-assisted context aware fertilizer recommendation." *IEEE Access* 10 (2022): 129505–129519.

10. Thorat, Tanmay, B. K. Patle, and Sunil Kumar Kashyap. "Intelligent insecticide and fertilizer recommendation system based on TPF-CNN for smart farming." *Smart Agricultural Technology* 3 (2023): 100114.

11. Swaminathan, Bhuvaneswari, et al. "IoT-driven artificial intelligence technique for fertilizer recommendation model." *IEEE Consumer Electronics Magazine* 12.2 (2022): 109–117.

12. Jain, Ravi Kant. "Experimental performance of smart IoT-enabled drip irrigation system using and controlled through web-based applications." *Smart Agricultural Technology* 4 (2023): 100215.

13. Tace, Youness, et al. "Smart irrigation system based on IoT and machine learning." *Energy Reports* 8 (2022): 1025–1036.

14. Ikram, Amna, et al. "Crop yield maximization using an IoT-based smart decision." *Journal of Sensors* 2022 (2022): 1–15.

15. Kitpo, Nuttakarn, and Masahiro Inoue. "Early rice disease detection and position mapping system using drone and IoT architecture." *2018 12th South East Asian Technical University Consortium (SEATUC)*, Vol. 1. IEEE, 2018.

16. Prathibha, S. R., Anupama Hongal, and M. P. Jyothi. "IoT-based monitoring system in smart agriculture." *2017 International Conference on Recent Advances in Electronics and Communication Technology (ICRAECT)*. IEEE, 2017.

17. Dlodlo, Nomusa, and Josephat Kalezhi. "The internet of things in agriculture for sustainable rural development." *2015 International Conference on Emerging Trends in Networks and Computer Communications (ETNCC)*. IEEE, 2015.

18. Khoa, Tran Anh, et al. "Smart agriculture using IoT multi-sensors: A novel watering management system." *Journal of Sensor and Actuator Networks* 8.3 (2019): 45.

19. Vishwajith, K. P., et al. "Decision support system for fertilizer recommendation – A case study." *Indian Journal of Agronomy* 59.2 (2014): 344–349.

20. Musanase, Christine, et al. "Data-driven analysis and machine learning-based crop and fertilizer recommendation system for revolutionizing farming practices." *Agriculture* 13.11 (2023): 2141.

21. Solorio-Ramírez, José-Luis, et al. "Random forest Algorithm for the Classification of Spectral Data of Astronomical Objects." *Algorithms* 16.6 (2023): 293.

3 Crop Yield Prediction Using Machine Learning Random Forest Algorithm

Suwarna Gothane

3.1 INTRODUCTION

According to the report of the Branch of Agriculture, collaboration, and Rancher Government Assistance Yearly Report 2018–19, farming assumes a unique part in India's economy. 54.6% Of the staff takes part in agrarian and associated area exercises and holds monetary records for 17.1% of the nation's Gross Worth. The issues of Indian ranchers are principally due to the small and divided land property. Quality seeds, excrements, composts, biocides, water systems, absence of automation, soil disintegration, and shortage of capital are realized issues related to the horticultural industry.

Initiation of good productivity of crops is possible by soil testing and educating farmers about fertilizers and organic farming. This helps to reduce soil erosion. A big step is required to educate farmers to avoid soil erosion and increase harvests using suitable fertilizers with advanced technology. One more solution to the agricultural sector is the need for agricultural marketing by means of recent knowledge or mobile phone applications.

3.2 LITERATURE SURVEY

John Havlin et al. proposed a solution for increasing crop productivity. The authors studied soil fertility management and increased crop nutrients and crop recovery on various plant management factors [1].

Oliver Knowles and Aimee Dawson worked on fertilizer and lime to increase crop yields and reduce loss of soil environment. In New Zealand, the authors carried out sampling processes to analyze soil nutrient characteristics [2]. Rebecca L et al. worked on non-linear correlation analysis with (NFIR) model and studied the properties of

DOI: 10.1201/9781003484608-3

ten types of soil on crop yield with comparison to the non-linear Random Forest (RF) model in Germany. Results were noticed in different cropping seasons, weather conditions, and crops [3].

Camila Fritzen Cidó investigated the relationship between organic agriculture and the bio-economy by carrying out a literature review. The author presented benefits to organic farmers in maintaining sustainability and green innovation [4].

Aneta Łukowska et al. proved soil testing is vital to protect the environment to increase production, and helps to save money and energy. The authors created a versatile stage with a soil testing gadget for agriculture, capacity of soil tests, and its investigation. The gathered examples gave exact information on supplements providing limits with expansion under productivity [5].

Chen Yunping et al. constructed a smart system for soil sampling. Five modules have been set up to respond to soil testing in view of a geographic data framework (GIS) and Global Positioning System (GPS) innovation utilizing flexible calculation with route elements and sensors. The designed system was analyzed and found effective for samples in the field [6].

Vrushali C. Waikar et al. proposed a system to create a model that helps farmers know which crop should be taken in a particular type of soil. The model suggested soil type, and according to soil type, it suggested suitable crops. Machine learning-based classifiers were used to suggest the crop [7].

3.3   PROPOSED WORK

The proposed work focuses on better crop yield and overcoming farmer problems. The prediction module predicts the harvest yield with the selected attributes from the informational collections. The model suggested crop yield using machine learning algorithms, with accuracy and a mobile app. It involves machine learning to pre-process data, feature extraction, and the use of an RF algorithm to determine crop yield. The architecture of crop yield prediction is shown in Figure 3.1.

This module returns the anticipated generation of crops based on the user's input. In case the client needs to know the generation of a specific crop, the framework takes the crop as the input, and it returns a list of crops at the side of their generation as yield. Actualized Steps:

1.1 Step 1: Select the functionality, i.e., crop prediction or yield forecast
1.2 Step 2: If the user chooses crop prediction, take soil type and area as input values, which are given as input to the RF implementation in the backend, and the corresponding predictions are returned. The algorithm returns a list of crops along with their predicted production.
1.3 Step 3: If the user chooses yield prediction: -Take crop, soil type, and area as inputs. These values are given as input to the RF implementation in the backend and the corresponding crop yield prediction I returned. The algorithm returns the predicted production of the given crop. The proposed system is divided into three modules: Data Collection and Analysis, K-Means Training Module, and Module to Perform Recommendation

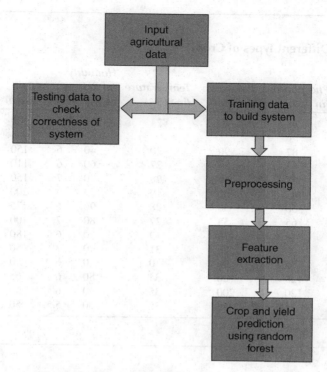

FIGURE 3.1   Architecture of crop yield prediction.

3.4   IMPLEMENTATION

We used Python 3.7 to identify different types of crops and yields. We have considered attributes including nitrogen range (60 – 300 kg/ha), phosphorus range (10 – 120 ppm), potassium range (40 – 200 ppm), temperature range (18 – 35°C/ degree Celsius), humidity range (10 – 90 RH value), pH range (5.5 – 7.0), rain range (30 – 250 cm). Based on the range values, different types of crops will be predicted (Tables 3.1 and 3.2).

3.5   CONCLUSION

In this research, we performed a survey and study of various systems for yield prediction. We proposed a feasible yield prediction and recommender system to the farmers, using the mobile application with features for crop prediction, selection, and planning using RF. Further results were compared with LSTM, RMM, Linear Regression, and K-Nearest Neighbor (KNN). Algorithms were tested, and results were compared for accuracy detection. Our approach using machine learning RF leads to an accuracy of 96%, which is observed to be greater than the mentioned algorithm. In the future, more accurate crop predictions and fertilizer usage for crop recommendation systems are required with statistical analysis.

TABLE 3.1
Results of Different Types of Crops

Nitrogen (in kg/ha)	Phosphorous (in ppm)	Potassium (in ppm)	Temperature (in °C)	Humidity (in RH value)	pH	Rain (inch)	Crop
64	48	40	24	25	5.5	100	Kidney beans
80	87	90	30	40	6	150	Coffee
72	69	84	27	60	6	110	Maize
70	78	89	28	90	7	150	Banana
110	45	100	19	25	6	200	Chickpea
140	50	50	28	90	6	175	Jute
128	65	55	22	80	7	100	Cotton
245	48	110	19	90	6.5	180	Rice
90	40	56	31	90	7	50	Mung bean
68	89	96	30	30	6	30	Muskmelon
100	120	100	34	80	6	65	Grapes
300	120	200	35	90	6	90	Apple
90	11	56	35	90	5.5	50	Watermelon

TABLE 3.2
Results of Different Types of Yield

State Name	Season	Crop Name	Area (acres)	Yield Prediction (sq/m)
Assam	Summer	Rice	6	5.17
Telangana	Winter	Rice	8	6.65
AP	Rainy	Tomato	2	0.99
Bihar	Rainy	Potato	5	5.20
Karnataka	Summer	Cotton	4	1.98
Kerala	Summer	Banana	3	1.26
Maharashtra	Rainy	Wheat	17	7.95
Maharashtra	Summer	Bajra	6	5.17
Jharkhand	Rainy	Chickpea	7	6.18
Jharkhand	Rainy	Paddy	10	7.14
Telangana	Summer	Mango	20	11.6
AP	Rainy	Cotton	3	1.26
Gujarat	Groundnut	Rainy	4.9	5.20
Gujarat	Cumin	Rainy	2.7	1.17
Chattisgarh	Summer	Sunflower	16	9.58
Tamil Nadu	Rainy	Coffee	12	5.92
Telangana	Rainy	Guava	11	7.69
Uttar Pradesh	Wheat	Summer	8	6.65
Punjab	Wheat	Rainy	5	5.2
Haryana	Wheat	Winter	9	6.44
Madhya Pradesh	Summer	Bengal gram	12	5.92

REFERENCES

1. John Havlin, Ron Heiniger: "Soil Fertility Management for Better Crop Production", MDPI, Agronomy, Volume: 10, 2020.
2. Oliver Knowles, Aimee Dawson: "Current Soil Sampling Methods – A Review", In: Farm Environmental Planning – Science, Policy and Practice (Eds. L. D. Currie and C. L. Christensen), Occasional Report No. 31. Fertilizer and Lime Research Centre, Massey University, Palmerston North, New Zealand, 2018. http://flrc.massey.ac.nz/publi cations.html.
3. Rebecca L. Whetton, Yifan Zhao, Said Nawar, Abdul M. Mouzen: "Modelling the Influence of Soil Properties on Crop Yields Using a Non-Linear NFIR Model and Laboratory Data", MDPI, Soil Systems, Volume: 5, 2021.
4. Camila Fritzen Cidó, Paola Schmitt Figueiró, Dusan Schreiber: "Benefits of Organic Agriculture Under the Perspective of the Bioeconomy: A Systematic Review", MDPI, Sustainability, Volume: 13 Issue: 12, 2021.
5. Aneta Łukowska, Piotr Tomaszuk, "Kazimierz Dzierżek, Łukasz Magnuszewski: Soil Sampling Mobile Platform for Agriculture 4.0", 20th International Carpathian Control Conference (ICCC), IEEE, 2019.
6. Chen Yunping, Wang Xiu, Zhao Chunjiang, "A Soil Sampling Intelligent System Based on Elastic Algorithm and GIS", Fifth International Conference on Natural Computation, IEEE Xplore, 28 December 2009.
7. Vrushali C. Waikar, Sheetal Y. Thorat, Ashlesha A. Ghute, Priya P. Rajput, Mahesh S. Shinde, "Crop Prediction Based on Soil Classification Using Machine Learning with Classifier Ensembling", *International Research Journal of Engineering and Technology*, Volume: 07 Issue: 05, 2020.

4 A Multi-Objective Based Genetic Approach for Increasing Crop Yield on Sustainable Farming

Swadhin Kumar Barisal, Gayatri Nayak, Bijayini Mohanty, Pushkar Kishore, Santilata Champati, and Alakananda Tripathy

4.1 INTRODUCTION

Sustainable agriculture is farming in a way that protects the environment, helps to grow natural resources, and makes the best use of non-renewable resources. It is also defined as ecosystem management, which involves the intricate interactions of soil, water, plants, animals, climate, and humans. To incorporate these aspects into farm-specific production strategies suited for the environment, people, and economic situations. Farms become ecologically sustainable by constructing a farm landscape resembling healthy ecosystems' complexity. Nature tends to work in cycles, with waste from one process or system becoming input for another. In contrast, industrial agriculture operates sequentially, like a factory: inputs enter one end, while products and trash exit. The other industrial and agricultural wastes (non-point-source pollution) include suspended dirt, nitrates and phosphates in streams, as well as nitrates and pesticides in groundwater. Sustainable agriculture is founded on the idea that a farm is a natural system rather than a factory. The easier we strive to make agriculture, the more exposed we are to natural calamities and market fluctuations [1,4–5]. We take a significant risk when we attempt to manufacture just one crop, such as wheat, corn, or soybeans. Instead, by diversifying crops and integrating plant and animal husbandry, expenses may be shared over several firms, lowering risk and improving profit.

Sustainable agricultural approaches strive to reduce farming's negative influence on the environment. It focuses on soil health, water conservation, and biodiversity protection [7]. Sustainable agriculture provides solutions to increasing worries about climate change and environmental deterioration. As the world's population grows, so does the need for natural resources like water, land, and energy [5–9]. Sustainable agricultural approaches improve resource efficiency, reduce waste, and ensure that resources are accessible for future generations. Sustainable agriculture

DOI: 10.1201/9781003484608-4

encourages farming methods adaptable to changing environmental circumstances and problems, including droughts, floods, and pests. Sustainable agriculture helps to ensure food security by diversifying crops, enhancing soil fertility, and applying water-saving practices. Sustainable agricultural strategies can increase the long-term economic viability of farming businesses [8]. Sustainable agriculture assists farmers in maintaining productivity while minimizing production risks and boosting market resilience by promoting soil conservation, lowering input costs, and enhancing biodiversity. Sustainable agricultural principles place a premium on social fairness, encouraging fair labor standards, supporting local communities, and ensuring small-scale farmers and underprivileged groups have equal access to resources [10–12]. Sustainable methods help to improve social well-being and rural development by promoting inclusive and equitable agricultural systems.

Resource allocation in sustainable agriculture is critical to the long-term viability of agricultural techniques and the conservation of natural resources. Sustainable agriculture seeks to fulfill present demands while preserving future generations' ability to meet their own. Effective resource allocation is critical to accomplishing this aim because it maximizes the utilization of diverse resources while reducing negative environmental consequences. The responsible management of water resources is an important part of resource allocation in sustainable agriculture [7–12]. Water shortage is a developing worldwide problem, worsened by climate change and unsustainable agricultural practices. Proper water resource allocation through drip irrigation, rainwater collection, and efficient water management strategies contribute to water conservation and ecosystem health. Similarly, land resource distribution is crucial for long-term agricultural sustainability [6–8]. As urbanization and land degradation encroach on arable land, there is an urgent need to prioritize the protection and long-term utilization of accessible land for agricultural purposes. Crop rotation, agroforestry, and soil conservation approaches can help maximize land use efficiency while improving soil health and fertility. Furthermore, allocating financial resources to sustainable agricultural practices is critical for promoting research, innovation, and the use of environmentally friendly technology. Investments in sustainable agriculture can increase crop yields, lower input costs, and more excellent resistance to environmental stresses. In addition to natural resources, human capital must be carefully allocated to sustainable agriculture [9–11]. Educating farmers about sustainable farming techniques, offering training programs, and fostering information exchange all help to establish resilient agricultural communities that can adjust to changing environmental conditions.

Optimization algorithms play a crucial role in resource allocation for sustainable agriculture, and their necessity stems from the complexities and challenges inherent in managing agricultural resources efficiently while balancing ecological, economic, and social considerations. Firstly, sustainable agriculture aims to maximize resource use efficiency while minimizing negative environmental impacts. Optimization algorithms offer a systematic approach to allocating resources such as water, fertilizers, and land, ensuring they are used optimally to achieve maximum yield with minimal environmental degradation. By optimizing resource allocation, farmers can reduce waste, conserve natural resources, and mitigate environmental pollution, contributing to long-term sustainability [8–13]. Secondly, agriculture operates within dynamic and uncertain environments influenced by factors like climate variability, market fluctuations, and changing consumer preferences. Optimization algorithms provide the flexibility to adapt resource allocation strategies in response to changing conditions, enabling farmers to optimize productivity and profitability while minimizing risks. These algorithms can incorporate real-time data and predictive models to make informed decisions, enhancing resilience and adaptive capacity in the face of uncertainties [15–17]. Furthermore, optimization algorithms help address the complexities of balancing multiple objectives in agricultural production, such as maximizing yield, minimizing input costs, and enhancing environmental stewardship [2–3]. Multi-objective optimization techniques allow farmers to explore trade-offs and find Pareto-optimal solutions that achieve optimal performance across conflicting objectives, promoting a holistic approach to sustainable agriculture [18–21].

A Genetic Optimization algorithm plays a crucial role in resource allocation for sustainable agriculture, providing a powerful and efficient approach to address complex optimization problems inherent in this domain. Sustainable agriculture involves balancing the need for increased food production with environmental preservation and resource efficiency [14]. The multifaceted nature of resource allocation in agriculture, considering factors such as land use, water usage, and crop selection, poses a challenging combinatorial optimization problem. Genetic Optimization algorithms excel in exploring large solution spaces, making them well-suited for finding optimal resource allocations that balance ecological sustainability and productivity [12–16]. The dynamic and evolving nature of agricultural systems requires adaptive optimization strategies. Genetic algorithms [22] [23] inherently support adaptability through mechanisms like crossover and mutation, allowing them to respond to changing environmental conditions, market demands, or technological advancements. This adaptability is crucial for sustainable agriculture, where flexibility is essential to cope with uncertainties in climate patterns, market dynamics, and emerging agricultural practices. Furthermore, Genetic Optimization algorithms facilitate the exploration of diverse and innovative resource allocation strategies. These algorithms can discover non-intuitive solutions that may escape traditional optimization approaches by leveraging principles inspired by natural selection and genetic variation [20]. This capability is vital in sustainable agriculture, where finding novel resource allocation schemes can lead to more resilient and environmentally friendly farming practices.

From the above discussion, we found that our objective in agriculture is facing increasing pressure to fulfill the rising global food demand while minimizing environmental effects. There is an urgent need for efficient resource allocation techniques in sustainable agricultural practices. Traditional agrarian systems frequently need more resource inefficiencies, resulting in waste, reduced yields, and environmental deterioration. Using Mathematical models allows you to optimize resource allocation and improve the sustainability of agricultural techniques. Aims to create and deploy optimal algorithms and machine learning models for successfully allocating resources of sustainable farm systems.

This article is organized clearly and concisely, with sections focusing on basic concepts, proposed mathematical model validation, results and discussion, managerial implications, conclusion, and future scope. The article is organized as follows: Section 2 for basic ideation and working principle of genetic algorithms (GAs), Section 3 for the proposed models and their working principles with the algorithm, Section 4 for mathematical validation of the proposed model with an example, and Section 5 for the conclusion.

4.2   METHODOLOGY

Start by initializing a population of potential solutions. This population typically consists of randomly generated individuals. Evaluate the fitness function $f(x)$ for each individual x in the population. $f(x)$ assigns a fitness score to each individual representing its quality or suitability for the problem at hand. Continue iterating through the following steps until a termination condition is met, which could be convergence or reaching a maximum number of generations. Select parents from the population based on certain criteria, such as probability p, individual x, and the fitness value $F(x)$. Common selection methods include roulette wheel selection, tournament selection, or rank-based selection. Generate offspring by performing crossover operations on selected parents. This operation creates new individuals by combining genetic material from two parent individuals. This step results in generating new individuals x_1, x_2, x_1', x_2'' through crossover. Initialize a mutation operation to introduce diversity into the population. Mutation perturbs the genetic material of individuals to explore new regions of the search space. Apply mutation to the offspring population, introducing random changes to their genetic material with respect to the mutation parameters C_0 and the mutation probability distribution P(m). Evaluate the fitness of the offspring population generated through crossover and mutation. Select individuals from the combined population of parents and offspring to form the next generation. The selection process may include strategies like elitism to ensure that the best individuals are preserved for the next generation. Once the termination condition is met, return the best solution found during the optimization process. After obtaining the best solution, update the machine learning models based on real-time data. This step ensures that the models adapt to changes in the environment or problem domain. Repeat the optimization process for subsequent farming seasons or whenever new data becomes available.

4.3   PROPOSED MODEL

We proposed an Opti-Grow model as shown in Figure 4.1, aims to improve resource allocation for sustainable agricultural methods by utilizing optimization algorithms. The approach seeks to solve the difficulties in agricultural systems by combining multiple components. By collecting and preparing data, the model guarantees that suitable data is available for analysis and decision-making. Preprocessing datasets, including soil qualities, weather patterns, crop kinds, historical yield data, and environmental aspects. It also improves data quality and prepares inputs for algorithms, resulting in more accurate insights from the data. GAs allow the model to estimate crop yields, detect insect infestations, anticipate weather trends, and discover hidden

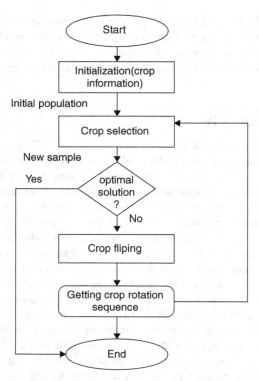

FIGURE 4.1   Proposed Opti-Grow model.

patterns in agricultural information. These insights enable farmers and others to make intelligent choices about crop management techniques and resource allocation strategies. To reduce waste and increase output, the optimization framework considers soil quality, water availability, nutrient levels, and sustainable development. The Opti-Grow model encourages resource efficiency and contributes to sustainable farming practices by optimizing resource allocation decisions. The decision support system enables farmers and other agricultural stakeholders to make intelligent choices about crop management techniques, resource allocation strategies, and risk mitigation measures. The solution improves farm production and sustainability by giving actionable insights in an easy-to-use manner.

To mathematically validate the approach, we must evaluate its theoretical underpinnings and the practicality of each step. In addition, we will use GAs for optimization, which is an excellent tool for handling complex optimization issues like resource allocation in agriculture.

4.3.1   INITIALIZATION OF RESOURCE ALLOCATION

Initial water, fertilizer, pesticide, and labor allocations are determined using historical data. This phase provides a starting point for resource optimization.

4.3.2 GENETIC ALGORITHM PSEUDOCODE

Here is a simplified pseudocode outlining the genetic algorithm for optimizing resource allocation in sustainable farming:

Algorithm 1: Opti-Grow: Genetic Algorithm for Sustainable Crop scheduling

---

Input: Crop information

Output: Crop Rotation Sequence

1. Initialize Crop Population P
2. Evaluate $f(x) \rightarrow X : R$ with respect to every P
3. max_generations = maximum generations
4. convergence = False
5. generation = 0
6. while not convergence and generation < max_generations: Select P(x) with respect to the p, x, $F(x)$
7. Generate Crop crossover followed by x_1, x_2, x', x''
8. Initialize m
9. Calculate m(C_o, P(m)) with respect to $P(m)$
10. Evaluate the fitness of the offspring Crop population
11. combined_population = combine (P, offspring_population)
    Select individuals from the combined population form the next generation
12. best_solution = find_best_solution(combined_population)
13. if best_solution meets termination condition:
14.   convergence = True
15.   generation += 1
16. Repeat the optimization process for subsequent farming seasons
17. Return the best solution found.

4.3.3 OPTIMIZATION USING GENETIC ALGORITHM

4.3.3.1 Initialization

Start by generating a population of potential solutions. Each solution is represented as a chromosome, typically a binary string or an array of values. Here, we have to create a population P of size N, where each individual represents a chromosome as per Eq. (4.1).

$$P = \{Chromosome1, Chromosome2, ..., ChromosomeN\} \qquad (4.1)$$

4.3.3.2 Objective Function

Define a fitness function, which quantifies how well each solution performs with respect to the problem we are trying to solve. This function is problem-specific and provides a numerical score for each chromosome. Eq. (4.2) shows the mathematical representation for a fitness function, denoted as $f(x)$, which quantifies how well each chromosome performs with respect to population P of size N, can be expressed as follows:

$$f(x) \to X : \mathbb{R} \tag{4.2}$$

Where:

- $f(x)$ *is the fitness function that takes a solution x as its input.*
- x *represents the space of potential solutions or chromosomes.*
- $\mathbb{R}$ *denotes the set of real numbers, and the output of the fitness function is a real numerical score that quantifies the performance of population P of size N.*

4.3.3.3 Selection

Choose individuals from the current population P to be parents of the next generation. The probability of selection is typically proportional to the fitness of the individuals. Here, we use the roulette wheel selection method for parent element selection. For the next generation, we select the parent element based on their fitness scores. Generally, the probability of selecting an individual is proportional to its fitness score. Eq. (4.3) represents the probability of selecting chromosome x as a parent using the roulette wheel selection method given by:

$$P(x) = \frac{f(x)}{\sum_{i=1}^{N} f(x)_i} \tag{4.3}$$

Where:

- N *is the total number of individuals in the population* P.
- *$f(x)$ is the fitness score of an individual with chromosome x.*
- *$P(x)$ is the probability of selecting an individual with chromosome x as a parent.*

4.3.3.4 Crossover (Recombination)

The process of creating new solutions (offspring) by combining the genetic information of two parent solutions in a genetic algorithm can be represented by using various crossover techniques. Here, we use the two-point crossover technique. Here we combine the genetic information of two parents to create offspring by using a crossover operator. From the two-parent solution, we have created two new offspring solutions. So, the crossover operator takes the genetic information of the two parents x_1 *and* x_2 and produces two offspring solutions, x' *and* x'' followed by the Eq. (4.4).

$$\mathbb{C}(x_1, x_2) = \left(x_1', x_2'' \right) \tag{4.4}$$

Where:

- x_1 and x_2 represents the two distinct parent solutions
- x_1' and x_2'' represents the offspring solutions.
- $\mathbb{C}(x_1, x_2)$ represents the crossover operator over the parent element x_1 and x_2.

4.3.3.5 Mutation

Apply mutation to some of the offspring to introduce small random changes in the chromosome. In GAs or genetic programming, the process of applying mutation to the offspring is a stochastic operation that introduces random changes to an individual's chromosome. Mutation helps maintain genetic diversity in the population. Here we have to apply a mutation operator with a small probability (0.001) to introduce random changes in the offspring's chromosomes. Eq. (4.5) represents the mathematical formulation for applying mutation with a probability to an offspring's chromosome can be described as:

$$C_n = m\left(C_{\vartheta}, P(m) \right)$$

(4.5)

Where:

- m represents the mutation.
- $P(m)$ as the probability of mutation 0.001.
- C chromosome as the representation of an individual's genetic information.
- C_o representants the old-chromosome.
- C_n representants the new-chromosome.
- $n(C_o, P(m))$ represents the mutation operation, which applies random changes to the with a C_o probability of $P(m)$.

4.3.3.6 Termination Criteria

The stopping criteria of an evolution process were defined based on their predefined conditions, such as the number of generations with a specific fitness threshold. Eq. (4.6) represents the conditional statement for a maximum number of generations and Eq. (4.7) represents the conditional statement for fitness of the best individual in the current generation with the fitness threshold.

The evolutionary process can be stopped when either of the following conditions is met:

$$\mathcal{G}_c \geq \mathcal{G}_{max}$$

(4.6)

$$\mathcal{B}\left[f(x) \right] \geq f(x)_T$$

(4.7)

Where:

- $\mathcal{G}_c$ represents the current generation

- $\mathcal{G}_{max}$ *represents* the maximum number of generations.
- $f(x)_T$ *represents the thresold value* T *with fittness value* $f(x)$.
- $B[f(x)]$ represents the best fitness values of the population P with $\mathcal{G}_c$.

4.4   WORKING EXAMPLE

We consider data with a set of crops that can be grown on a farm, each with specific requirements and benefits and limited research resources, and determine the
optimal allocation of research efforts to maximize the overall yield over a certain
period through crop rotation optimization. Let us consider a simplified scenario
with the following crops: Corn(C), Wheat(w), Soybean(S), Potato(P), Rice(R),
Oil crops(O), Bajra(B), Cotton (Co), Jowar(J), Mellite(M), Cauliflower (Ca),
Grams(G). The farmer has a limited research budget of 100 hours, and each crop
rotation plan requires a certain amount of research hours to evaluate its viability
and effectiveness. Additionally, each crop has specific benefits in yield improvement and soil health enhancement. Basically, our objective is to maximize the
overall yield and soil health enhancement while respecting the research budget
constraint.

After having all the parameters, we can formulate the optimization problem as
a mathematical model. Let us define the decision variables, objective function, and
their constraints.

   i. Decision Variables:
      Let x_i represent the binary decision variable indicating whether to allocate
      research efforts to plan i or not. $x_i = 1$ if research is allocated to plan i, and
      $x_i = 0$ otherwise.

  ii. Constraints:
      For our problem statement we have to consider two different types of constrains:

4.4.1   BUDGET CONSTRAINT

The total allocated research hours cannot exceed the budget.

$$\sum_{i=1}^{n}(\text{Hours}_i \times x_i) \leq 100 \tag{4.8}$$

Where,

- *Hours $_i$ represents the research hours required for Plan i.*
- x_i *is the binary decision variable indicating whether to allocate research
  efforts to Plan i or not.*

Explanation:

- Multiply the research hours required for each plan $Hours_i$ by the binary decision variable x_i to represent whether the research is allocated to that plan.
- The total ensures that this sum does not exceed the research budget of 100 hours.

4.4.2 BINARY VARIABLE CONSTRAINT

Here, x_i represents the binary variable Constraints. The value of x_i belongs to $\{0, 1\}$ where, $i = 1,2\ldots..12$.

$$x_i \in \{0, 1\} \ \forall \ i = (1, 2 \ldots \ldots .12) \qquad (4.9)$$

iii. Objective Function:

The objective function aims to maximize the overall yield improvement and soil health enhancement while considering the selected crop rotation plans.

$$Maximize \ Z = \sum_{i=1}^{n}\left[\left(15\% \times Yield_i\right)+\left(5\% \times soil_i\right)\right]\times x_i \qquad (4.10)$$

- $15\% \ Yield_i$ represents the yield improvement percentage for crop i.
- $5\% \ Soil$ represents the soil health enhancement percentage for crop i.
- X_i is the binary decision variable indicating whether to allocate research efforts to Plan i.

Explanation of the objective function:

- For each crop rotation plan i, calculate the total benefit considering both yield improvement and soil health enhancement.
- Multiply the yield improvement percentage of crop i by 15% and the soil health enhancement percentage of crop i by 5%.
- Multiply the total by the binary decision variable x_i to represent whether this plan is selected or not.

iv. Mathematical Model:

We formulate our mathematical model followed by Eq. (4.8 – 4.10).

$$Maximize \ Z = \sum_{i=1}^{n}\left[\left(15\% \times Yield_i\right)+\left(5\% \times soil_i\right)\right]\times x_i$$

$$subject \ to \sum_{i=1}^{n}\left(Hours_i \times x_i\right) \leq 100$$

$$x_i \in \{0,1\} \ \forall \ i = 1,2,\ldots.12$$

Let us solve the optimization problem to find the optimal allocation of research efforts. Let us expand the model using Table 4.1 & Table 4.2. We get the value of Z as follows:

TABLE 4.1
Represents the Profit and Soil Health Parameters with Their Duration

Crop Name	Yield Improvement (Profit)	Soil Health Enhance	Duration Required (Hour)
Corn(C)	15%	5%	2
Wheat(w)	10%	7%	7
Soybean(S)	12%	6%	8
Potato(P)	8%	9%	6
Rice(R)	15%	6%	9
Oil Crops(O)	35%	3%	5
Bajra(B)	11%	4%	4
Cotton (Co)	17%	1%	10
Jowar(J)	13%	4%	7
Mellite(M)	40%	8%	2
Cauliflower (Ca)	31%	2%	4
Grams(G)	23%	8%	5

TABLE 4.2
Represents Crop Rotation Plan

Plan Name	Crop Rotation Plan	Plan Name	Crop Rotation Plan
Plan 1	C- W-S-P-B-O-G-R-J	Plan 11	P-C-W-G-R-J -G-M
Plan 2	C- W-S-P-R-J-O-Co	Plan 12	O-R-W-R-P-O-Co
Plan 3	S-P-C-W-G-M-O-B	Plan 13	P-C-W-R-M-Co-G
Plan 4	P-C-W-S- G-M-O-B	Plan 14	R-J -G-M- P-C-W-G
Plan 5	W-G-M-O-B -P-C-Co	Plan 15	R-W-C-J-G-M-Co
Plan 6	G-M-O-B-R-J-W-M	Plan 16	R-P-O-Co-J-W-G
Plan 7	R-J -G-M-B -P-C-Co	Plan 17	P-Ca-J-W-G-B-R-C
Plan 8	W-R-J -G-M-O-B	Plan 18	R-J-P-C-W-G-M-Co
Plan 9	P-C-W-G-M-O-B	Plan 19	J-W-G-B-R-P-O-Co
Plan 10	W-G-M-Co-J-R-Co	Plan 20	R-Ca-J-P-C-W-G

$$Z = [(0.15 \cdot Yield_1) + (0.05 \cdot Soil_1)]x_1 + [(0.15 \cdot Yield_2) + (0.07 \cdot Soil_2)]x_2$$
$$+ [(0.15 \cdot Yield_3) + (0.06 \cdot Soil_3)]x_3 + (0.15 \cdot Yield_4) + (0.09 \cdot Soil_4)x_4$$
$$+ [(0.15 \cdot Yield_5) + (0.05 \cdot Soil_5)]x_5 + (0.15 \cdot Yield_6) + (0.07 \cdot Soil_6)x_6$$
$$+ [(0.15 \cdot Yield_7) + (0.06 \cdot Soil_7)]x_7 + (0.15 \cdot Yield_8) + (0.09 \cdot Soil_8)x_8$$

$$+ [(0.15 \cdot Yield_9) + (0.05 \cdot Soil_9)]x_9 + [(0.15 \cdot Yield_{10})$$
$$+ (0.07 \cdot Soil_{10})]x_{10} + [(0.15 \cdot Yield_{11}) + (0.06 \cdot Soil_{11})]x_{11}$$
$$+ [(0.15 \cdot Yield_{12}) + (0.09 \cdot Soil_{12})]x_{12}$$

Subject to: $20x_1 + 25x_2 + 30x_3 + 35x_4 + 20x_5 + 26x_6 + 15x_7 + 33x_8 + 35x_9 + 30x_{10} + 28x_{11} + 40x_{12} \le 100$

$$x_1, x_2, x_3, \ldots \ldots \ldots x_{12} \in \{0,1\}$$

Solution:

Primarily we have to find the optimal solution for $x_1, x_2, x_3, \ldots \ldots \ldots x_{12}$ and then calculate the value of *Maximize Z*.

Let's start solving:

$$20x_1 + 25x_2 + 30x_3 + 35x_4 + 20x_5 + 26x_6 + 15x_7 + 33x_8 + 35x_9 +$$
$$30x_{10} + 28x_{11} + 40x_{12} \le 100 \ldots \ldots \ldots \ldots \quad (11)$$

Given that each x_i can only be 0 or 1, we have to maximize the value of Z while satisfying the constraint.

After solving Eq. (11), we got our optimal values as:

$$x_1 = 1, x_2 = 1, x_3 = 0, x_4 = 1, x_5 = 1, x_6 = 1, x_7 = 0, x_8 = 1, x_9 = 1, x_{10} = 1, x_{11} = 0, x_{12} = 1$$

Now, substitute these values into the objective function:

$$Z = [(0.15 \cdot Yield_1) + (0.05 \cdot Soil_1)]x_1 + [(0.15 \cdot Yield_2) + (0.07 \cdot Soil_2)]x_2$$
$$+ [(0.15 \cdot Yield_3) + (0.06 \cdot Soil_3)]x_3 + (0.15 \cdot Yield_4) + (0.09 \cdot Soil_4)x_4$$
$$+ [(0.15 \cdot Yield_5) + (0.05 \cdot Soil_5)]x_5 + (0.15 \cdot Yield_6) + (0.07 \cdot Soil_6)x_6$$
$$+ [(0.15 \cdot Yield_7) + (0.06 \cdot Soil_7)]x_7 + (0.15 \cdot Yield_8) + (0.09 \cdot Soil_8)x_8$$
$$+ [(0.15 \cdot Yield_9) + (0.05 \cdot Soil_9)]x_9 + [(0.15 \cdot Yield_{10}) + (0.07 \cdot Soil_{10})$$
$$+ [(0.15 \cdot Yield_{11}) + (0.06 \cdot Soil_{11})]x_{11} + (0.15 \cdot Yield_{12})$$
$$+ (0.09 \cdot Soil_{12})x_{12}$$

Then we have to substitute the values of yield improvement and soil health enhancement. We get,

$$Z = [0.15 \times 15\% + 0.05 \times 5\%] + [0.15 \times 10\% + 0.05 \times 7\%]$$
$$+ [0.15 \times 8\% + 0.05 \times 9\%] + [0.15 \times 15\% + 0.05 \times 6\%]$$
$$+ [0.15 \times 35\% + 0.05 \times 3\%] + [0.15 \times 17\% + 0.05 \times 1\%]$$
$$+ [0.15 \times 13\% + 0.05 \times 4\%] + [0.15 \times 40\% + 0.05 \times 8\%]$$
$$+ [0.15 \times 23\% + 0.05 \times 8\%]$$

Let us solve for Z

$$
\begin{aligned}
Z ={}& (0.15 \times 0.15) + (0.05 \times 0.05)] + {}^{|}(0.15 \times 0.10) + (0.05 \times 0.07)^{|} \\
&+ {}^{|}(0.15 \times 0.08) + (0.05 \times 0.09)^{|} + {}^{|}(0.15 \times 0.15) + (0.05 \times 0.06)^{|} \\
&+ {}^{|}(0.15 \times 0.35) + (0.05 \times 0.03)^{|} + (0.15 \times 0.17) + (0.05 \times 0.01) \\
&+ (0.15 \times 0.13) + (0.05 \times 0.04) + (0.15 \times 0.40) + (0.05 \times 0.08) \\
&+ (0.15 \times 0.23) + (0.05 \times 0.08) \\
\Rightarrow Z ={}& 0.0225 + 0.0025 + 0.015 + 0.0035 + 0.0525 + 0.0045 + 0.0225 \\
&+ 0.0030 + 0.0525 + 0.0015 + 0.0255 + 0.0005 + 0.0195 \\
&+ 0.0020 + 0.0060 + 0.0040 + 0.0345 + 0.0040
\end{aligned}
$$

$$\Rightarrow Z = 1.18$$

So, *the Maximize Z* is approximately 1.18.

v. Validation of Working Example:

The implementation of a crop rotation strategy that corresponds to the best alloca-tion of research resources. For better analysis, we tested our model in python 3.11.3 software. We found the best crop rotation plan by GA approach with maintaining the yield and soil health enhanced. So, the chosen crop rotation plans are in the following order: P-C-W-R-M-Co-G or a similar rotation pattern. Allocate resources like man-power, machinery, fertilizers, and water in accordance with the crop rotation plan you have set. Irrigation, soil preparation, insect management, and other particular requirements may vary by crop. Figure 4.2 represents the yield improvement and soil health enhancement parameters with their duration.

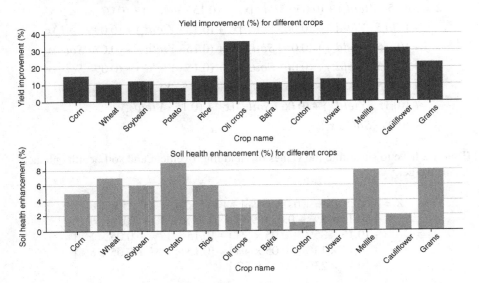

FIGURE 4.2   Yield improvement (%) for different crops and soil health enhancement (%) for different crops.

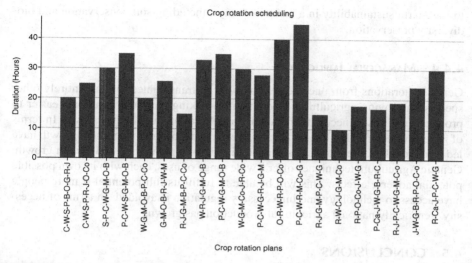

FIGURE 4.3 Crop rotation plans for best farming.

Figure 4.3 represents the crop rotation plans for best farming. Allocate resources efficiently to ensure that each crop obtains the inputs it needs for maximum development and output. Keep track of variables like growth rate, pest and disease incidence, soil health indicators, and weather conditions. Adjust resource allocation and management procedures as needed to address problems or improve crop production. At the end of the growing season, assess the effectiveness of the specified crop rotation strategy. Measure each crop's actual yield and evaluate soil health indicators such as organic matter content, nitrogen levels, and microbial activity. Compare the actual results to the projected outcomes using the optimization model.

4.4.3 RESULT & DISCUSSION

Our mathematical approach determines the optimal resource allocation, which attempts to maximize both crop improvement and soil health. We should expect major advantages in agricultural productivity and soil fertility if research efforts are directed toward specific crop rotation strategies. The crop rotation plans used have been adjusted to increase production by systematically rotating crops with complimentary growth patterns and nutritional requirements. This can result in greater crop yields over time by reducing soil depletion and better utilizing available resources. Farmers should expect a significant increase in crop yield when research efforts focus on these optimal strategies, as opposed to traditional farming approaches. Crop rotation has been shown to promote soil health by minimizing pests, illnesses, and nutrient depletion. Rotating crops with varying nutrient demands allows the soil to restore key nutrients organically, resulting in better soil structure, fertility, and microbial diversity. The focus of research efforts on crop rotation plans that prioritize soil health enhancement demonstrates a commitment to sustainable agricultural practices and long-term soil conservation. The chosen crop rotation plans contribute to long-term sustainability in a variety of ways, including soil conservation and bio-diversity preservation.

4.4.4 MANAGERIAL IMPLICATION

Genetic alterations from one species can now be transplanted into an entirely other species, improving agricultural production or making particular materials easier to produce. In terms of necessity, national agricultural production has grown. In terms of genetic effects, technological advancements, and slower consumption growth have led to national progress toward sustainability despite population and wealth growth. Genetic Agriculture techniques offer a wide range of uses, each with its own possible potential adverse effects. This will be able to fulfill its full potential if more people have access to technology and unnecessary regulation is avoided. In terms of necessity, genetically national agricultural production has grown.

4.5 CONCLUSIONS

In conclusion, mathematical models and optimization algorithms offer a potential method for improving resource allocation efficiency and increasing agricultural sustainability. Farmers can use these technologies to make more informed decisions and enhance resource utilization in their farming operations. The framework is a systematic strategy that provides farmers with tools and technologies to maximize the advantages of their agricultural activities while avoiding resource waste and environmental concerns. Farmers can utilize mathematical models to establish optimal farming practices by analyzing complicated relationships between several aspects, such as crop rotation, soil health, water availability, and climate. Farmers can use optimization algorithms to determine the most efficient allocation of resources, such as manpower, machinery, fertilizers, and water, to meet their goals. This allows farmers to increase crop yields, improve soil health, and reduce inputs, lowering costs and environmental impact. As agriculture evolves, complex technologies such as mathematical modeling and optimization algorithms are predicted to become increasingly important in maintaining food security, environmental stewardship, and socioeconomic development in farming communities. Farmers that embrace and use these techniques in their farming methods can adapt to changing environmental circumstances, reduce risks, and strengthen the resilience of their agricultural systems. Mathematical models and optimization algorithms offer a powerful solution for enhancing resource allocation efficiency and sustainability in agriculture. By integrating these tools into farming practices, farmers can optimize their operations, increase productivity, and contribute to a more sustainable and resilient agricultural sector.

REFERENCES

1.   Ali, A., Hussain, T., Tantashutikun, N., Hussain, N., & Cocetta, G. (2023). Application of smart techniques, internet of things and data mining for resource use efficient and sustainable crop production. *Agriculture, 13*(2), 397.

2. Barisal, S. K., Chauhan, S. P. S., Dutta, A., Godboley, S., Sahoo, B., & Mohapatra, D. P. (2022). BOOMPizer: Minimization and prioritization of CONCOLIC based boosted MC/DC test cases. *Journal of King Saud University – Computer and Information Sciences, 34*(10), 9757–9776.

3. Barisal, S. K., Dutta, A., Godboley, S., Sahoo, B., & Mohapatra, D. P. (2021). MC/DC guided test sequence prioritization using firefly algorithm. *Evolutionary Intelligence, 14*, 105–118.

4. Bhattacharjee, P., & Bhattacharya, S. (2023). Increasing annual profit of wind farm using improved genetic algorithm. *International Journal of Advanced Natural Sciences and Engineering Researches, 7*(4), 203–209.

5. Brito, L. F., Bédère, N., Douhard, F., Oliveira, H. R., Arnal, M., Peñagaricano, F., ... & Miglior, F. (2021). Genetic selection of high-yielding dairy cattle toward sustainable farming systems in a rapidly changing world. Animal, *15*, 100292.

6. Cao, K., Huang, B., Wang, S., & Lin, H. (2012). Sustainable land use optimization using Boundary-Based Fast Genetic Algorithm. *Computers, Environment and Urban Systems, 36*(3), 257–269.

7. Cheng, D., Yao, Y., Liu, R., Li, X., Guan, B., & Yu, F. (2023). Precision agriculture management based on a surrogate model assisted multiobjective algorithmic framework. *Scientific Reports, 13*(1), 1142.

8. Das, V., & Jain, S. (2020). Genetic algorithm to find most optimum growing technique for multiple cropping using big data. In *Emerging Technologies for Agriculture and Environment: Select Proceedings of ITsFEW 2018* (pp. 77–94). Springer Singapore.

9. Fathollahi-Fard, A. M., Tian, G., Ke, H., Fu, Y., & Wong, K. Y. (2023). Efficient multi-objective metaheuristic algorithm for sustainable harvest planning problem. *Computers & Operations Research, 158*, 106304.

10. Monica, N. I., Pooja, S. R., Rithiga, S., & Madhumathi, R. (2023, March). Soil NPK prediction using enhanced genetic algorithm. In *2023 9th International Conference on Advanced Computing and Communication Systems (ICACCS)* (Vol. 1, pp. 2014–2018). IEEE.

11. Karamian, F., Mirakzadeh, A. A., & Azari, A. (2023). Application of multi-objective genetic algorithm for optimal combination of resources to achieve sustainable agriculture based on the water-energy-food nexus framework. *Science of The Total Environment, 860*, 160419.

12. Khoshnevisan, B., Bolandnazar, E., Shamshirband, S., Shariati, H. M., Anuar, N. B., & Kiah, M. L. M. (2015). Decreasing environmental impacts of cropping systems using life cycle assessment (LCA) and multi-objective genetic algorithm. *Journal of Cleaner Production, 86*, 67–77.

13. Komi, M. A., Tonnang, H. E., Elfatih, M., Odindi, J., Mutanga, O., & Saliou, N. (2023). Data-driven Artificial Intelligence (AI) Algorithms for modelling potential maize yield under maize–legume farming systems in East Africa. Agronomy, *12*(12), 3085. https://doi.org/10.3390/agronomy12123085

14. Liu, Y., Jiang, C., Lu, C., Wang, Z., & Che, W. (2023). Increasing the accuracy of soil nutrient prediction by improving genetic algorithm backpropagation neural networks. *Symmetry, 15*(1), 151.

15. Ma, X. (2023). Smart agriculture and rural revitalization and development based on the internet of things under the background of big data. *Sustainability, 15*(4), 3352.

16. Mishra, A., & Goel, L. (2023). Metaheuristic algorithms in smart farming: An analytical survey. *IETE Technical Review, 41*(4), 1–20.

17. Naeem, K., Zghibi, A., Elomri, A., Mazzoni, A., & Triki, C. (2023). A literature review on system dynamics modeling for sustainable management of water supply and demand. *Sustainability*, *15*(8), 6826.

18. Nayak, G., Barisal, S. K., & Ray, M. (2023). CGWO: An Improved Grey Wolf Optimization Technique for Test Case Prioritization. *Programming and Computer Software*, *49*(8), 942–953.

19. Olakulehin, O. J., & Omidiora, E. O. (2014). A genetic algorithm approach to maximize crop yields and sustain soil fertility. *Net Journal of Agricultural Science*, *2*(3), 94–103.

20. Schöning, J., Wachter, P., & Trautz, D. (2023). Crop rotation and management tools for every farmer: The current status on crop rotation and management tools for enabling sustainable agriculture worldwide. *Smart Agricultural Technology, 3*, 100086.

21. Srivastava, S., & Yagysen, D. (2016). Implementation of genetic algorithm for agriculture system. *International Journal of New Innovations in Engineering and Technology*, *5*(1), 82–86.

22. Xu, T., Yao, L., Xu, L., Chen, Q., & Yang, Z. (2023). Image segmentation of cucumber seedlings based on genetic algorithm. *Sustainability*, *15*(4), 3089.

23. Zhou, Y., & Fan, H. (2018). Research on multi-objective optimization model of sustainable agriculture industrial structure based on genetic algorithm. *Journal of Intelligent & Fuzzy Systems*, *35*(3), 2901–2907.

5 Drones for Crop Monitoring and Analysis

Preethi Nanjundan, Indu P.V., and Lijo Thomas

5.1 INTRODUCTION

Drone technology integration for crop monitoring and analysis has caused a paradigm change in the agriculture sector in recent years. Unmanned aerial vehicles (UAVs), commonly referred to as drones, have become very potent instruments that provide farmers with previously unheard-of possibilities to enhance crop management strategies and raise total yield. The purpose of this introduction is to give a general overview of the importance of drones in contemporary agriculture, emphasizing how they are transforming crop monitoring and analysis. Farmers have always depended on satellite imaging and ground-based techniques to monitor crop health and spot any problems. However, these methods frequently fall short of the precision, promptness, and economy needed to satisfy the needs of contemporary agriculture. Drones provide a versatile and effective way to gather high-resolution data across these restrictions. Drones that are outfitted with sophisticated sensors, such as thermal and multispectral cameras, may gather comprehensive data on insect infestations, nutrient shortfalls, crop health, and other important indicators. To provide farmers with useful insights, this data can be processed by specialist software or examined in real time. Drones enable farmers to carry out tailored treatments, such as variable rate fertilization, precision irrigation, and targeted spraying, therefore maximizing resource utilization and minimizing environmental effects. This is made possible by the early and precise detection of issues. Drones also provide farmers with unmatched accessibility and flexibility, enabling them to continuously monitor their crops and modify their management plans as needed throughout the growing season. Rapid aerial crop condition assessment offers an all-encompassing viewpoint that is not achievable with ground-based techniques alone. Drones have, all things considered, completely changed crop monitoring and analysis by giving farmers rapid, accurate, and useful information that they can use to improve agricultural sustainability and profitability. Drone technology is predicted to play an ever-bigger role in agriculture as it develops, opening up new avenues for precision farming and guaranteeing food security in a world that is constantly changing.

5.2 OVERVIEW OF THE ROLE OF DRONES IN AGRICULTURE

In contemporary agriculture, drones have become indispensable instruments, offering farmers an unprecedented array of capabilities for crop management, surveillance,

DOI: 10.1201/9781003484608-5

and analysis. These UAVs, outfitted with advanced sensors and cameras, provide farmers with high-resolution data and photos so they can make well-informed decisions all through the growing season [1]. One of the most notable uses of drones in agriculture is precision farming. Drones provide focused treatments like variable rate fertilization, precision irrigation, and localized pest management by providing specific insights about crop health, moisture levels, nutrient deficits [2], and insect infestations. This accuracy minimizes the impact on the environment while maximizing the use of resources, increasing agricultural production. Drones not only enable precision agriculture but also offer a cost-efficient way to monitor crops over large agricultural areas. Drones often record overhead imagery thanks to their sophisticated sensors, which enable farmers to quickly detect and handle problems like disease outbreaks, [3] weed infestations, and water stress. The proactive monitoring of crops leads to increased resilience and production on farms. Drones are also essential for field mapping and surveying since they provide accurate maps and three-dimensional (3D) representations of farms. With this knowledge of soil variability, topography, and drainage patterns, farmers can plan their property [4], choose crops, and build infrastructure with confidence. Farm profitability and efficiency are therefore greatly increased. Drones also aid in environmental stewardship by assessing soil erosion, plant cover, and water quality. Drones provide farmers with the ability to enforce compliance and use sustainable land management practices by collecting data [5] on important environmental indicators in accordance with legal requirements. This proactive monitoring strategy for the environment promotes long-term agricultural sustainability [6].

5.2.1 IMPORTANCE OF CROP MONITORING AND ANALYSIS FOR MODERN FARMING PRACTICES

Modern agricultural techniques heavily rely on crop monitoring and analysis, which provides farmers with vital information on the health, growth, and yield of their crops. The need for efficient and effective agricultural monitoring and analysis is critical, given the rising demand for food production to support the world's expanding population. In this paragraph, we shall go to great lengths on the importance of crop monitoring and analysis for contemporary agricultural methods. Crop monitoring and analysis, which are fundamental to contemporary agriculture, allow farmers to make well-informed decisions from planting to harvest. Farmers who constantly monitor their crops are better able to evaluate the health of the plants, spot possible problems, and take prompt action to maximize yields and maintain crop quality. To maximize yield, this proactive approach to crop management is crucial while reducing losses brought on by illnesses, pests, and environmental [7] stresses. Farmers can also obtain useful information from crop monitoring and analysis, which helps them make decisions about a variety of farm management issues. Farmers may adjust their cultivation techniques to meet the unique requirements of their crops and maximize resource allocation by gathering data on variables like soil moisture, nutrient levels, and insect populations. For instance, accurate irrigation scheduling based on real-time moisture data may assist in avoiding overwatering and saving water, and tailored fertilization techniques can guarantee that crops get the nutrients they require without using excessive amounts of pesticides. Moreover, crop monitoring and analysis are essential to sustainable agriculture because they help farmers limit their impact on the environment and save expenses associated with inputs. By precisely evaluating, farmers can implement integrated pest management (IPM) techniques, which give priority to biological treatments and reduce the use of chemical pesticides [8], to address crop health and pest challenges. Similarly, farmers may decrease the danger of water contamination and reduce nutrient runoff by adjusting fertilizer applications depending on crop requirements and soil nutrient levels. Crop monitoring and analysis help enhance decision-making in areas including crop selection, rotation, and land use planning, in addition to enhancing crop management techniques. Farmers are able to choose which crops to sow in a given season by evaluating past data on crop performance and environmental factors, including soil compatibility, market demand, and climate resilience. Similar to this, farmers may spot trends and patterns by tracking crop performance over time [9]. Farmers can implement IPM techniques, which give priority to biological treatments and reduce the use of chemical pesticides, to address crop health and pest challenges. Similarly, farmers may decrease the danger of water contamination and reduce nutrient runoff by adjusting fertilizer applications depending on crop requirements and soil nutrient levels. Crop monitoring and analysis help enhance decision-making in areas including crop selection, rotation, and land use planning, in addition to enhancing crop management techniques. Farmers are able to choose which crops to sow in a given season by evaluating past data on crop performance and environmental factors, including soil compatibility, market demand, and climate resilience. Similar to this, farmers may spot trends and patterns by tracking crop performance over time. With the use of technology, farmers may optimize their operations' sustainability, profitability, and production by [10] making well-informed decisions. Crop monitoring and analysis will continue to be essential instruments for balancing the needs of an expanding population with the preservation of our natural resources for future generations, even as the difficulties facing agriculture change over time.

5.2.2 APPLICATIONS OF DRONES IN CROP MONITORING

Drone applications for crop monitoring have transformed contemporary agriculture by giving farmers previously unheard-of capacity to precisely and effectively manage their farms. With their sophisticated sensors and cameras, these UAVs provide a wide range of applications that improve crop monitoring methods and enable farmers to make informed decisions during the growing season. Precision agriculture is one of the main uses of drones for crop monitoring [11]. Drones provide farmers with exceptional precision in assessing crop health, moisture levels, nutritional shortages, and insect infestations by recording high-resolution imagery and data. With the use of this data, customized treatments that maximize agricultural yields and optimize resource use while reducing environmental effects are made possible, such as variable rate fertilization, precision irrigation, and localized insect control. Apart from precision farming, UAVs play a vital part in the evaluation of crop health. Drones with specialized sensors are able to recognize early signs of disease outbreaks, weed infestations, and nutrient deficits, as well as minor changes in crop health. Early identification enables farmers to prevent losses and maintain crop health in a timely manner, eventually resulting in a more [12] strong and resilient yield. Drones are also excellent for surveying and mapping fields, giving farmers access to precise maps and three-dimensional (3D) representations of their property. These maps provide important information on topography, drainage patterns, and soil variability, which helps with land planning, crop selection, and infrastructure development decision-making. Farmers may increase overall farm production and efficiency by improving land usage and cultivation techniques based on this geographical data. Another crucial aspect is environmental monitoring. Application of drones in agricultural monitoring. Drones can evaluate plant cover, soil erosion, and water quality, giving farmers useful information to assist sustainable land management techniques [13]. Drones assist farmers in minimizing their environmental effects and adhering to regulations by tracking environmental indicators, thus assuring the long-term sustainability of their businesses. Drones are also essential for estimating crop yields. Drones can predict agricultural yields accurately by analyzing data and aerial photography, which helps farmers plan logistics, anticipate harvests, and make strategic marketing decisions. Farmers may increase profitability and streamline harvest processes with this understanding of crop production, which benefits their agricultural operations as a whole. In conclusion, drones have a multitude of uses for crop monitoring, giving farmers access to real-time information. Providing useful information to guarantee environmental sustainability, enhance production, and optimize crop management techniques. Drone technology is projected to play an ever bigger role in agriculture [14] as it develops, spurring efficiency and innovation in the farming sector. Utilizing drone technology, farmers can guarantee food security in a world that is always changing and open up new opportunities for precision farming.

5.3 AERIAL IMAGING FOR CROP HEALTH ASSESSMENT

One essential use of drones in contemporary agriculture is aerial photography for crop health evaluation, which provides farmers with unmatched crop condition data. Drones with specialized sensors and cameras are able to take high-resolution pictures of fields, which facilitates in-depth examination of crop health markers, including biomass, color, and canopy cover. By identifying regions of stress or disease outbreaks that may not be visible from ground level, farmers may rapidly and correctly evaluate the general health and vigor of their crops using this airborne viewpoint. Drones' capacity to successfully and economically cover vast agricultural regions is one of the main benefits of employing them to check crop health. It takes much time and work to monitor crops using traditional techniques like driving or walking around fields. It is challenging to conduct thorough and routine crop monitoring. Drones, on the other hand, can quickly and easily gather comprehensive imagery by flying over fields at different elevations [1]. Frequent monitoring enables farmers to maximize yields and minimize crop losses by identifying problems early and taking proactive efforts to rectify them.

Drones also provide an unparalleled degree of accuracy and detail compared to other aerial photography methods, such as satellite images. Drones can now take pictures at extremely high resolutions thanks to developments in sensor technology, which makes it possible to precisely analyze agricultural health indicators down to the plant level. Accurately evaluating crop health and seeing minute variations in growth and development that can point to underlying problems need this degree of information. Additionally, the information gathered by drones may be automatically processed and analyzed [2] using sophisticated software algorithms to identify abnormalities and crop health problems. By using patterns and features found in the picture, machine learning algorithms may be trained to recognize certain crop illnesses or insect infestations, facilitating quick and precise issue identification. Farmers are better able to recognize and address crop health concerns promptly thanks to this computerized analysis, which expedites the decision-making process. All things considered, aerial imagery for crop health monitoring is a potent use of drone technology that provides farmers with a host of advantages. Drones give farmers the ability to make well-informed decisions that improve crop management techniques, increase yields, [3] and protect the long-term health of their crops by giving them precise, fast, and thorough information on the state of their crops. as well as the longevity of their business. The potential for aerial imagery in agriculture is endless as drone technology develops, offering even more efficacy and efficiency in crop monitoring and management.

5.3.1 IDENTIFICATION OF PEST INFESTATIONS AND DISEASE OUTBREAKS

Another critical use of drones in crop monitoring is the detection of pest infestations and disease outbreaks, which gives farmers a proactive approach to managing pests and diseases. Drones with specialized sensors [4] and cameras are able to take high-resolution pictures of fields, which makes it possible to identify pests and illnesses early on before they pose a threat to the health of crops. Drones can swiftly and effectively cover enormous agricultural regions, which is one of the main benefits of employing them to identify pests and diseases. It can be difficult to consistently and thoroughly monitor crops using labor-intensive, time-consuming traditional techniques of scouting for [5] pests and diseases, such as driving or walking across fields. Drones, on the other hand, are able to fly over fields at different heights and take precise pictures of crops from above. This view from above enables farmers to scan vast tracts of land far faster than they could using ground-based techniques.

Drones also provide an unparalleled degree of accuracy and detail compared to other aerial photography methods [6], such as satellite images. Drones with advanced sensor technology are able to take pictures at extremely high resolutions, which makes it possible to precisely identify pests and illnesses at the plant level. With this level of information, farmers may see early symptoms of illness or infestation like discoloration, wilting, or insect damage before they spread widely and result in large crop losses. Moreover, sophisticated image processing and machine [7] learning algorithms may be used to evaluate the data gathered by drones, allowing for the automatic identification of disease outbreaks and pest infestations. Through instruction by using algorithms to identify certain patterns and traits linked to pests and illnesses, farmers are able to promptly detect and react to new threats in real time. Farmers may carry out targeted interventions, like pesticide treatments or crop rotations, to reduce the spread of pests and diseases and safeguard agricultural yields since this automated analysis expedites [9] the decision-making process. Drone detection of disease outbreaks and pest infestations is, all things considered, a potent weapon in the farmer's toolbox that allows for early detection and quick action in the event of a danger to crop yield and health. Farmers may minimize crop losses, use proactive pest management tactics, and guarantee the long-term viability of their enterprises by utilizing drones for pest and disease monitoring. As drone technology and agriculture advance, the potential for identifying pests and diseases is endless, offering even higher levels of efficiency and efficacy in crop management and protection [10].

5.3.2 MONITORING CROP GROWTH AND YIELD ESTIMATION

In order to ensure food security and sustainable farming practices, modern agriculture relies heavily on monitoring crop growth and yield estimation. This process entails the systematic observation, measurement, and analysis of various factors influencing crop development, resulting in accurate predictions of potential yields. In this essay, we will discuss the significance of crop growth monitoring and the methods used for yield estimation. First, monitoring crop growth is necessary for timely interventions and the optimization of agricultural practices. Farmers and agronomists use a variety of techniques. These techniques include remote sensing. Farmers can detect problems like insect infestations, illnesses, or nutritional deficits by regularly monitoring these data, which enables them to take timely and focused corrective action [11]. The method we use to track crop development has changed dramatically as a result of advances in remote sensing technology like satellite imaging and drones. Large agricultural expanses may be seen from above with the help of these instruments, which also provide important information about crop health and growth trends. High-resolution photographs may be taken by drones with sophisticated sensors, which can be used to identify abnormalities or early indicators of stress in the field. On the other hand, regular and extensive monitoring made possible by satellite imaging enables a more thorough understanding of crop growth at the regional or even global level. Precise yield estimation is essential for efficient farm management and planning, in addition to monitoring. Forecasting the potential harvest assists farmers in making well-informed choices about marketing tactics, resource allocation, and risk management in general. To estimate yields, a variety of models and algorithms are used, taking [13] into account variables including soil quality, weather, and historical data. These forecasts are now much more accurate because of machine learning and artificial intelligence, which analyze enormous databases and spot intricate patterns. Crop models are one technique often used to estimate yield. Based on inputs, including meteorological data, soil properties, and crop management techniques, these models mimic the growth and development of crops. Farmers and academics may simulate various scenarios and predict the effects of variable changes on crop yields by incorporating these components. In the end, this proactive strategy facilitates improved decision-making and resource use. Crop development makes proactive interventions possible, which enhances yields and promotes sustainable [14] farming methods and resource efficiency. The means by which we can track and forecast crop growth will become increasingly important as agricultural technology advances in order to meet the demands of feeding a growing world population.

5.3.3 Assessing Soil Health and Fertility

In order to ensure the productivity and resilience of farmland for future generations, sustainable agriculture requires an assessment of the health and fertility of the soil. Often called the "living skin of the Earth," soil is a complex ecosystem that is rich in organic matter, nutrients, and bacteria. This essay will discuss the value of evaluating the fertility and health of the soil as well as the instruments and techniques employed in this process. First and foremost, the ability of soil to support biodiversity, grow plants, and preserve ecosystem services is a key indicator of soil health [1]. A balanced organic matter content, sufficient nutrition levels, sound structure, and biological activity are the hallmarks of healthy soil. By evaluating soil health, farmers may put management strategies into place and comprehend the state of their soil at the moment. to preserve or raise its quality. Measuring soil fertility is a crucial component in evaluating soil health. The capacity of the soil to supply vital nutrients to plants in the proper amount and balance is referred to as soil fertility. For plants to grow and develop, several nutrients are essential, including micronutrients, phosphorus, potassium, and nitrogen [2]. Farmers may utilize soil testing to assist them in making well-informed decisions about applying fertilizer and adding soil amendments by finding out the pH and nitrogen levels of the soil. Gathering soil samples from different parts of a field and examining them for pH, nutrients, and other characteristics is known as soil testing. Farmers may learn important information about the nutrient level and any inadequacies of the soil via laboratory examination. These findings allow farmers to create specialized fertilization programs that are suited to the particular requirements of their soil types and crops. Soil health evaluation includes a variety of physical, chemical, and biological markers in addition to nutrition analysis. Soil structure, porosity, and other physical characteristics affect root growth, water infiltration, and erosion susceptibility [3]. Microbial activity and nutrient availability are impacted by chemical characteristics such as pH, cation exchange capability, and organic matter concentration. Microbial biomass, earthworm activity, and enzyme activity are examples of biological indicators that show the total biological variety and health of the soil ecosystem. Technological developments have resulted in the creation of novel instruments and methods for evaluating the fertility and health of soil. For example, real-time monitoring of soil temperature, moisture content, and nutrient levels is made possible using soil sensors. In the field, these sensors can be used to gather ongoing data to help farmers decide when to use fertilizer, water, and other management techniques. Furthermore, the geographical distribution of characteristics and soil variability may be better understood with the use of digital soil mapping and modeling approaches [4]. Through the integration of data obtained from soil surveys, remote sensing, and geographic information systems

(GIS), farmers are able to generate intricate maps that depict the characteristics and variations in their farms' soil. Decisions about crop selection, planting density, and nutrient management techniques may all be made more effectively with the use of this information.

5.3.4 DATA ANALYSIS TECHNIQUES

The foundation of contemporary data-driven decision-making processes in a variety of disciplines and sectors is data analysis methodologies. These procedures involve a wide range of techniques meant to extract insights, patterns, and trends from unprocessed data. Using efficient data analysis tools is crucial for producing actionable insights and promoting well-informed decision-making in the big data age when information is plentiful but sometimes overwhelming. Because they summarize the key characteristics of a dataset, descriptive statistics provide the basis of data analysis. A quick glance at the central tendency and dispersion of the data is provided by metrics [1] like mean, median, mode, standard deviation, and range. Before going further into analysis, analysts may better grasp the fundamental qualities of the data by using these statistics, which provide preliminary insights into its features. With inferential statistics, data taking analysis is a step further by utilizing a sample of data to draw conclusions or predictions about the population. Analysts are able to deduce linkages and patterns in the data by using techniques like regression analysis, analysis of variance (ANOVA), and hypothesis testing. Through the process of extrapolating data from a sample to a larger population, inferential statistics offer significant insights into wider patterns and trends. With its visual method of data analysis, exploratory data analysis (EDA) enhances descriptive and inferential statistics. With the use of graphical representations like histograms, scatter plots, box plots, and heat maps, EDA entails analyzing a dataset's key features. Through the identification of patterns, outliers, and links in the data, these visualizations aid in the development of hypotheses and subsequent investigation. Statistical models are utilized in predictive analytics and a dataset while keeping all of its pertinent data. Techniques that convert [7] high-dimensional data into lower-dimensional representations, such as principal component analysis (PCA) and t-distributed stochastic neighbor embedding (t-SNE), make analysis and visualization easier. Reducing the number of dimensions in a dataset makes it easier to handle and understand. Unstructured text data is analyzed using text mining and Natural Language Processing (NLP) techniques in order to identify trends and insights. Text data from sources like social media, consumer reviews, and documents is processed and analyzed using techniques like named entity recognition (NER), sentiment analysis, and topic modeling. Organizations may improve decision-making and consumer engagement by using text mining and NLP to extract meaningful insights from large volumes of unstructured text data. The goal of time-series analysis is to find patterns, trends, and seasonality in data that has been gathered over an extended period of time. Methods Time-series data are analyzed and forecast using techniques including Fourier analysis, exponential smoothing, and autoregressive integrated moving average (ARIMA) modeling. In order to forecast future trends and make well-informed judgments based on previous data, time-series analysis is frequently utilized in finance, economics, and other disciplines.

5.4 IMAGE PROCESSING AND ANALYSIS FOR CROP HEALTH ASSESSMENT

Modern agriculture now uses image processing and analysis as essential tools, providing creative approaches to crop health evaluation. Farmers and researchers may monitor crops on a broad scale, identify early symptoms of stress or illness, and enhance agricultural methods by using photos' visual information. The importance of image processing and analysis in crop health assessment, as well as the techniques and uses that make it a useful tool in precision agriculture, will all be covered in this article. Crop health evaluation is one of the main uses of image processing in agriculture [8]. Drones and satellites are examples of remote sensing technology that can take high-resolution pictures of agricultural areas. These photos offer insightful information on a number of crop health-related topics, including potential stresses, nutritional levels, and plant vigor. By removing relevant information from these images, image processing techniques improve the visuals and make it possible to judge crop conditions in more depth and accuracy. The capacity of image processing to identify minute alterations in plant traits that might not be apparent to the unaided eye is a significant benefit in the evaluation of crop health. For example, variations in the size, color, or texture of leaves might be signs of illnesses, insect infestations, or nutritional deficits. These alterations may be detected by sophisticated picture analysis algorithms, which enable farmers to take preventive action to solve issues before they have a major influence on crop productivity.

Spectral imaging is another effective method for evaluating crop health. Spectral cameras use several electromagnetic spectrum bands to record pictures, giving comprehensive details on crops' reflectance characteristics. Numerous vegetation indices, including the Normalized Difference Vegetation Index (NDVI), which measures the quantity of chlorophyll in plants, may be calculated using this data. NDVI is frequently used to evaluate the general health of crops and pinpoint specific regions in the field that could need care. Image processing helps with precision agriculture by enabling specific treatments in addition to monitoring the general health of the crops. Farmers can use image analysis to pinpoint certain parts of a field that need different irrigation, fertilizer, or pest management strategies. This focused strategy reduces the negative environmental effects of agricultural operations while simultaneously increasing resource efficiency. Artificial intelligence (AI) and machine learning are essential for automating image processing for crop health evaluation. These technologies facilitate the creation of models that are trained to identify abnormalities and patterns in pictures. To categorize fresh photographs and detect possible problems, machine learning algorithms, for instance, can be trained on a [7] library of photos with known crop health concerns. Large agricultural regions may now be swiftly and precisely analyzed thanks to this automation, which also increases crop monitoring's scalability and efficiency. Furthermore, the timely completion of crop health evaluations is improved by the integration of real-time data and image processing. Identification of dynamic changes in the field, such as the quick spread of illnesses or abrupt changes in climatic circumstances, is made possible by continuous monitoring. Farmers may minimize losses and stop problems from getting worse by detecting problems early and taking appropriate action.

5.4.1 MACHINE LEARNING ALGORITHMS FOR AUTOMATED DETECTION OF CROP ANOMALIES

The automatic identification of crop abnormalities made possible by machine learning algorithms has revolutionized the way farmers monitor and manage their farms. Machine learning algorithms can detect minute variations in crop health by utilizing large datasets and sophisticated computational methods. This allows for the early identification of pests, illnesses, nutrient shortages, and other problems. This essay will examine the role that machine learning plays in automated agricultural anomaly detection, as well as the techniques and uses that make it an invaluable tool for precision farming. The speed and accuracy with which machine learning algorithms can evaluate enormous datasets is one of its main advantages in agricultural anomaly identification. Conventional anomaly detection techniques can be laborious since they frequently rely on manual examination or present criteria. and prone to mistakes. In contrast, machine learning algorithms have the capacity to precisely detect patterns and abnormalities by analyzing enormous volumes of data from a variety of sources, such as sensor data [6], drone footage, and satellite photography. Crop anomaly detection often makes use of supervised learning methods like regression and classification. When using supervised learning, the system is trained using a labeled dataset that includes instances of both typical and unusual crop circumstances. The system is able to generalize its knowledge and correctly categorize new occurrences of anomalies in unseen data by learning from these examples. In order to enable preventive management techniques, a classifier trained on photos of both healthy and diseased plants, for instance, may automatically recognize indicators of illness in new photographs.

Unsupervised learning strategies, such as grouping and crop anomaly detection, as well as anomaly detection. The algorithm's job in unsupervised learning is to find patterns and abnormalities in unlabeled data without being aware of what constitutes normal or abnormal circumstances beforehand. Algorithms for clustering can be used to discover regions of the field that have similar traits by grouping together comparable cases of crop health. On the other hand, anomaly detection systems highlight occurrences that markedly depart from the average, suggesting possible problems [5] that need more research. Convolutional Neural Networks (CNNs) have shown to be especially successful machine learning models for crop anomaly identification based on images. CNNs are deep learning architectures that are ideal for tasks like object identification and image classification since they are made to automatically build hierarchical representations of pictures. Using extensive datasets of cropped photos with annotations, researchers trained CNNs can create models with high accuracy that can identify a variety of irregularities, including nutrient deficits and insect infestations. Crop anomaly detection also makes use of Long Short-Term Memory (LSTM) networks and Recurrent Neural Networks (RNNs), especially for time-series data. Because RNNs and LSTMs can model sequential data and capture temporal relationships, they are perfect for assessing sensor data that is gathered over an extended period of time. Researchers can forecast future crop abnormalities based on patterns found in the data by training these models on previous sensor data. This allows for pre-emptive management techniques to reduce possible dangers. Additionally, the efficiency and scalability of agricultural anomaly detection are improved by the integration of machine learning algorithms with remote sensing technologies, such as satellites and drones. These resources offer high-definition photos of agricultural fields, allowing for a thorough examination of crop health across vast geographic areas. Farmers can rapidly and accurately monitor large agricultural regions by merging machine learning algorithms with remote sensing data. This allows for early interventions to maintain crop quality and output.

5.4.2 CHALLENGES AND FUTURE DIRECTIONS

The landscape of precision agriculture and environmental monitoring is being shaped by a number of difficulties and emerging approaches in the field of combining drone data with GIS for spatial analysis [1]. The intricacy of handling and evaluating massive amounts of drone images inside GIS platforms is one major obstacle. Drones may collect precise information on natural landscapes or agricultural fields in high-resolution, but the sheer volume of data they collect can be too much for typical GIS systems to handle. The creation of scalable and effective data processing methods, such as distributed data storage and parallel computing, is necessary to meet this problem. Furthermore, the process of analyzing drone data inside GIS settings may be streamlined and made faster and more accurate by integrating machine learning algorithms for automatic feature extraction and categorization. The compatibility and interoperability of drone systems with GIS applications is another difficulty. Drones can output data in different formats, resolutions, and coordinate systems, [5] which might make it difficult to integrate them seamlessly with GIS applications. Drone data integration with GIS platforms may be facilitated and interoperability may be improved by standardizing data formats and metadata for drone photography. Additionally, cooperation and innovation in the field of drone-based spatial analysis may be fostered by creating open-source software tools and APIs (Application Programming Interfaces) for data integration and interchange. Another major obstacle to the integration of drone data with GIS for spatial analysis is data quality assurance and accuracy. Data produced by drones can be inaccurate or unreliable depending on a number of factors, including air conditions, flying settings, and sensor calibration. Putting strict quality control in place is vital to implement strategies like ground trothing and validation against reference data to guarantee the precision and [12] dependability of spatial analysis outcomes. Moreover, including inertial navigation systems (INS) and real-time GPS (Global Positioning System) corrections can improve the accuracy of drone data collecting, enhancing the caliber of spatial analysis results inside GIS platforms. [3] The integration of drone data with GIS for spatial analysis presents extra problems because of privacy and data security concerns, especially in sensitive or restricted locations. Collecting, processing, and sharing drone-derived data in GIS contexts raises serious challenges for protecting sensitive data and guaranteeing adherence to data privacy laws. Robust encryption, access control, and anonymization strategies may be put into practice to protect private data and lessen privacy threats related to drone-based spatial analysis. Additionally, creating drone technology in GIS applications can be more widely trusted and accepted if clear, moral standards are followed for data gathering and use. Future prospects for integrating drone data with GIS for spatial analysis show promise in a number of areas. Multispectral and [13] hyperspectral imaging are examples of advances in sensor technology that provide new possibilities for obtaining rich spatial data on crops, vegetation, and environmental factors. By combining these cutting-edge sensors with drones, we can precisely and thoroughly monitor wetlands, forests, farms, and other natural ecosystems, which improves our comprehension of intricate spatial dynamics and processes.

5.5 CONCLUSION

In summary, even if there are obstacles to overcome, the integration of drone data with GIS has great potential for spatial analysis. Due to the enormous amount of drone data and pictures, effective processing techniques and compatibility with current GIS systems are needed. Clear parameters are required for drone operations and data processing due to regulatory compliance and privacy issues. Future developments in sensor capabilities, data processing algorithms, and drone technology will improve the efficacy of drone-based spatial analysis. New perspectives and uses will be made possible by integration with cutting-edge technology like machine learning and AI. The combination of drones with GIS presents previously unheard-of possibilities for sustainable management and well-informed decision-making in a variety of industries despite some obstacles. By resolving issues related to technology, regulations, and privacy, we can fully utilize drone data integration with GIS to tackle challenging spatial issues and promote creativity in geographic analysis.

REFERENCES

1. Balestro, G., Fioraso, G., & Lombardo, B. (2013). Geological map of the Monviso massif (Western Alps). *Journal of Maps*, 9(4), 623–634.
2. Chen, S. C., Hsiao, Y. S., & Chung, T. H. (2015). Determination of landslide and driftwood potentials by fixed-wing UAV-borne RGB and NIR images: a case study of Shenmu Area in Taiwan. In: EGU General Assembly Conference Abstracts (Vol. 17; p. 2491).
3. Dabove, P., Manzino, A. M., & Taglioretti, C. (2014). GNSS network products for post-processing positioning: limitations and peculiarities. *Applied Geomatics*, 6(1), 27–36.
4. Endres, F., Hess, J., Sturm, J., Cremers, D., & Burgard, W. (2013). 3-D mapping with an RGB-D camera. *IEEE Transactions on Robotics*, 30(1), 177–187.
5. Farfaglia, S., Lollino, G., Iaquinta, M., Sale, I., Catella, P., Martino, M., & Chiesa, S. (2015). The use of UAV to monitor and manage the territory: perspectives from the SMAT project. In Engineering Geology for Society and Territory – Volume 5 (pp. 691–695). Springer.

6. Giordan, D., Manconi, A., Tannant, D. D., & Allasia, P. (2015, July). UAV: low-cost remote sensing for high-resolution investigation of landslides. In 2015 IEEE International Geoscience and Remote Sensing Symposium (IGARSS) (pp. 5344–5347). IEEE.

7. Joyce, K. E., Samsonov, S. V., Levick, S. R., Engelbrecht, J., & Belliss, S. (2014). Mapping and monitoring geological hazards using optical, LiDAR, and synthetic aperture RADAR image data. *Natural Hazards*, 73(2), 137–163.

8. Mendes, T., Henriques, S., Catalao, J., Redweik, P., & Vieira, G. (2015, October). Photogrammetry with UAV's: quality assessment of open-source software for generation of ortophotos and digital surface models. In Proceedings of the VIII Conferencia Nacional De Cartografia e Geodesia, Lisbon, Portugal (pp. 29–30).

9. Sadeghipoor, Z., Lu, Y. M., & Su¨sstrunk, S. (2015, February). Gradient-based correction of chromatic aberration in the joint acquisition of color and near-infrared images. In *Digital Photography XI* (Vol. 9404, p. 94040F). International Society for Optics and Photonics.

10. Shi, B., & Liu, C. (2015, December). UAV for landslide mapping and deformation analysis. In *International Conference on Intelligent Earth Observing and Applications* 2015 (Vol. 9808, p. 98080P). International Society for Optics and Photonics.

11. Song, C., Yue, C., Zhang, W., Zhang, D., Hong, Z., & Meng, L. (2019). A remote sensing-based method for drought monitoring using the similarity between drought eigenvectors. *International Journal of Remote Sensing*, 40(23), 8838–8856.

12. Taddia, G., Gnavi, L., Piras, M., Forno, M. G., Lingua, A., & Russo, S. L. (2015). Landslide susceptibility zoning using GIS tools: an application in the Germanasca valley (NW Italy). In Engineering Geology for Society and Territory – Volume 2 (pp. 177–181). Springer.

13. Vasuki, Y., Holden, E. J., Kovesi, P., & Micklethwaite, S. (2014). Semi-automatic mapping of geological structures using UAV-based photogrammetric data: an image analysis approach. *Computers & Geosciences*, 69, 22–32.

14. Westoby, M. J., Brasington, J., Glasser, N. F., Hambrey, M. J., & Reynolds, J. M. (2012). 'Structure-from-Motion' photogrammetry: a low-cost, effective tool for geoscience applications. *Geomorphology*, 179, 300–314.

6 Decision Support System for Sustainable Farming

*Nandini Nenavath, Venkata Krishna Reddy M.,
and Chintala Sai Akshitha*

6.1 INTRODUCTION

India ranks among the top agricultural-producing nations in the world. India is a major producer of agricultural products, yet its farm productivity is still low. Farm productivity is low, which means that farmers make relatively little money. Increased production is required to provide farmers with a higher income. Farmers need to know which crop is best for a certain plot of land to get more yield. If the right kind of crop is planted on that piece of land, the yield will undoubtedly increase. Therefore, farmers can benefit greatly from crop recommendation systems. Crop growth is influenced by a variety of factors. The yield is dependent upon various parameters such as rainfall, temperature, pH, humidity, and the amount of potassium, phosphorus, and nitrogen present in the soil. Many farmers are unaware of the best crop to plant and the location in which to get the most yield. In this proposed system, parameters like rainfall, soil features, temperature, etc., are taken into consideration to predict suitable crops, thus increasing crop productivity.

6.2 RELATED WORK

Many research studies have been carried out in the agricultural sector to improve productivity in recent years. Rajesh, C. et al. [1] utilized a variety of supervised learning algorithms, such as Random Forest, Support Vector Machine, and Linear Regression, to analyze a dataset comprising 2200 instances with selected attributes such as soil nutrients (nitrogen, phosphorus, potassium), pH level, rainfall, temperature, and humidity. The results showed that Random Forest emerged as the top performer, achieving the highest accuracy. Additionally, the analysis of the dataset samples revealed that rice crops require higher nitrogen levels and significant rainfall, while pigeon peas demonstrated optimal growth in regions characterized by abundant rainfall. Overall, this investigation highlights the potential of machine learning algorithms and sensor data in improving crop yields, reducing resource waste, and increasing profitability in sustainable agriculture.

DOI: 10.1201/9781003484608-6

Motwani, A. et al. [2] implemented a Convolutional Neural Network (CNN) to effectively classify input images, distinguishing between various soil types, including red soil, black soil, alluvial soil, and clay soil. Building upon this classification, the study incorporated state-specific information and employed the Random Forest algorithm to predict appropriate crop types based on anticipated yields tailored to the specific soil characteristics identified earlier. Consequently, the study aimed to offer uniform crop recommendations to farmers sharing similar soil types within the same state, thus facilitating an efficient and effective decision-making framework for agricultural practices.

Jadhav, A. et al. [3] applied various supervised algorithms, including Random Forest, Decision Tree, and Logistic Regression, where Random Forest demonstrated superior performance, achieving the highest accuracy. The dataset under consideration comprises 2200 instances featuring selected attributes such as soil nutrients (nitrogen, phosphorus, potassium), pH level, rainfall, and cultivation area. However, it lacks essential environmental characteristics like temperature and humidity, which are crucial for understanding weather conditions.

Sagana, C. et al. [4] addresses the prevalent practice among farmers of prioritizing crops based solely on market value and potential profits, often neglecting crucial factors such as soil quality and sustainability, which can have detrimental effects on both the environment and the farmers themselves. To rectify this, the study harnesses the potential of modern machine learning techniques to provide effective crop recommendation support. Evaluating multiple models, including Random Forest, XG Boost, Naïve Bayes, Logistic Regression, Decision Tree, and Support Vector Machine. The research aims to guide farmers in making informed decisions that balance economic profitability with sustainable agricultural practices, ultimately promoting long-term environmental health and the well-being of farming communities.

Gayathiri, B. et al. [5] gathered their dataset from the Salem district of Tamil Nadu. Random Forest was utilized for crop recommendation, while K-means clustering was employed for fertilizer recommendation. The authors also suggested using fertilizer to enhance crop yield, taking into consideration the soil's nitrogen, phosphorus, and potassium (NPK) levels. Yield predictions were made for rice and maize for the entire year, providing valuable information to farmers about expected crop yields.

Jeevaganesh, R. et al. [6] not only recommended crops but also predicted their yield using the Ad- boost algorithm, which combines multiple weak classifiers to enhance accuracy. Simultaneously, fertilizer recommendations were made through the Random Forest algorithm. Notably, the crop recommendation system considered parameters such as rainfall, location, pH value, area, soil, humidity, temperature, season, and label, excluding soil characteristics. The system utilizes location as an input for crop recommendation, while fertilizer recommendation factors include soil type, crop type, humidity, temperature, and moisture.

Israni et al. [7] explored modern approaches to precision farming, which can help farmers predict crop yields and choose the right crops to cultivate. The parameters temperature, area, district, and rainfall were used to predict crop yields and recommend crops to farmers. Machine learning algorithms like Ridge Regression, XGB Regressor, and LGBM Classifier achieved better accuracy in predicting crop yields and recommending crops. The XGB Regressor with Hyperparameter tuning and LGBM Classifier with Hyperparameter tuning produced the best results in their models. A notification system should be implemented using SMS or email to inform farmers of the predicted crop yield and recommended crops.

Devdatta A. Bondre et al. [8] used supervised learning algorithms to predict crop yield and suggest suitable fertilizers for different crops, which is the main goal of this research. The study used soil samples from various areas of Jammu District and analyzed soil nutrients and environmental characteristics. Using different classifier algorithms, the problem of crop yield prediction was formulated as a classification task. On the basis of accuracy and execution time, performance evaluations of classification algorithms were conducted in comparison. The findings showed that crop yields for a variety of crops, including wheat, cotton, onions, jowar, sugarcane, rice, dry chili, and soybeans, can be predicted with the highest accuracy.

Macharla, Ranjith. et al. [9] proposed a machine learning-based system that helps farmers make informed decisions about which crops to plant based on various environmental factors. The system was designed to take into account factors such as soil type, temperature, humidity, and rainfall to provide recommendations on the optimal crop to plant. It is composed of three main modules: the Administration Module, the User Module, and the Machine Learning Model Module. An administrator is in charge of loading the model when the user requests it and training the models on the data. The input data, which includes rainfall, pH as well as nitrogen, phosphorus, and potassium (N-P-K) values, can be submitted by the user to the administrator. After being trained on the dataset, the machine learning model makes predictions using the data that is fed to it. It is designed to meet specific criteria, including accuracy, outperforming the present system, and being able to communicate with the existing system.

Mr. Omkar Kulkarni. et al. [10] present a novel approach to crop recommendation systems using supervised learning algorithms. The designed system helps farmers in India make knowledgeable choices about suitable crops based on soil conditions and other environmental characteristics. The system uses a web portal to collect soil details and other relevant information from farmers and then uses machine learning algorithms to predict which crop is suitable based on the user-given inputs. Highlights are the importance of precision farming and the need for modern tools and infrastructure to improve crop yield and promote sustainable agriculture. The proposed system is based on a Random Forest algorithm, which has been shown to be effective in generating accurate crop prediction results. The system also includes features like OTP-based authentication, location picker, and image recognition to enhance user experience and improve the accuracy of the predictions. The related work discussed above is tabulated in Table 6.1.

TABLE 6.1
Observations from Different Research Papers

S NO	Paper Title	Year	Methodology	Observations
1	Smart Crop Recommendation System Using Machine Learning	2023	Random Forest, Linear Regression, and Support Vector Machine	Achieving the highest accuracy using Random Forest helps in increasing the profitability of sustainable agriculture.
2	Soil Analysis and Crop Recommendation using Machine Learning	2022	Convolutional Neural Network (CNN)	Classifies input images, distinguishing between various soil types, including red soil, black soil, alluvial soil, and clay soil, resulting in recommendations for farmers with the same crop having similar soil types within the same state.
3	Crop Recommendation System Using Machine Learning Algorithms	2022	Random Forest, Logistic Regression and Decision Tree	The author did not consider the essential environmental characteristics like temperature and humidity, which are crucial for understanding weather conditions.
4	Machine Learning-Based Crop Recommendations for Precision Farming to Maximize Crop Yields	2023	Random Forest, XG Boost, Naïve Bayes, Logistic Regression, Decision Tree, and Support Vector Machine	The author only considers the market value and potential profits rather than the suitable crop for the soil.
5	Machine Learning-based Crop Suitability Prediction and Fertilizser Recommendation System	2023	Random Forest, K-means Clustering	The author also suggested using fertilizer to enhance crop yield, taking into consideration the soil's NPK levels.

(continued)

TABLE 6.1 (Continued)
Observations from Different Research Papers

S NO	Paper Title	Year	Methodology	Observations
6	A Machine Learning-based Approach for Crop Yield Prediction and Fertilizer Recommendation	2022	Adaboost algorithm, Random Forest	The author also predicted the yield using the Adaboost algorithm.
7	Crop yield prediction and crop recommendation system	2022	Ridge Regression, XGB Regressor, and LGBMClassifier	The author mainly used parameters based on the area for crop yield and recommend crops. Proposed notification system using SMS or email to inform farmers of the predicted crop yield.
8	Prediction of crop yield and fertilizer recommendation using machine learning algorithms	2019	Random Forest, Support Vector Machine	The author emphasises applying Supervised learning algorithms to determine crop yield and suggest appropriate fertilizers for particular crops.
9	Crop Recommendation System	2022	Random Forest, KNN neighbor, and Decision Tree	The administration, user, and machine learning model modules are the three mod- ules. The administrator is in charge of load- ing the model when the user requests it and training the models on the data. Upon receiving the user data, the administrator loads and presents the outcome.
10	Crop Recommendation System using Machine Learning algorithm	2022	Random Forest	The author used an OTP-based authentication and also image recognition, which helps to identify diseased leaves and the preventive measures to be followed.

6.3   PROPOSED METHODOLOGY

To design and develop an integrated system that utilizes comprehensive data on environmental and soil conditions and provides personalized advice for appropriate crop choices. Additionally, the system should provide insights into the appropriate use of fertilizers, considering soil nutrient levels and crop-specific requirements. The system also aims to incorporate an analysis of past pesticide usage, enabling farmers to make knowledgeable choices regarding pesticide management and minimizing the environmental impact of agricultural practices. The main objectives are:

- To enhance agricultural productivity and profitability by assisting farmers in making informed decisions about crop choices, thereby ensuring sustainable farming practices and maximizing yield.
- Optimize crop yield and minimize environmental impact from fertilizer recommendation

Methodology
- Data collection: Gather data on crops, environment, and fertilizers.
- Data preprocessing: Ensure uniformity and normalize data for analysis.
- Model development: Use supervised learning algorithms like Random Forest, Decision Tree, Logistic Regression, Naive Bayes, Support Vector Machine etc. to predict and recommend.
- Suitable crops: Based on environmental conditions (rainfall, humidity, pH, P, temperature, N, K).
- Fertilizer usage: Based on crop choice and soil conditions.
- Recommendation system integration: Integrate the trained model into a recommendation system and take input values based on its system recommendations.
- Interface development: Design user-friendly interfaces for crop and fertilizer recommendations. Methodology as shown in Figure 6.1.

6.4   IMPLEMENTATION

In the proposed system, 'Smart Farming', the below dataset for crop prediction, and Random Forest (RF) and K-Nearest Neighbors (KNN) machine learning algorithms are applied to the dataset.

6.4.1   DATASET

The dataset will enable users to build a predictive model that recommends which crops would be best to plant on a particular farm. The dataset shown in Figure 6.2 comprises 2200 instances, drawn from historical data spanning 22 distinct varieties of unique crops, including banana, coconut, papaya, orange, muskmelon, rice, jute, blackgram, watermelon, grapes, mango, chickpeas, mothbeans, maize, coffee, lentil, cotton, kidney beans, pomegranate, pigeonpeas, and apple and it is derived from sources such as kaggle with the soil characteristics and environmental characteristics

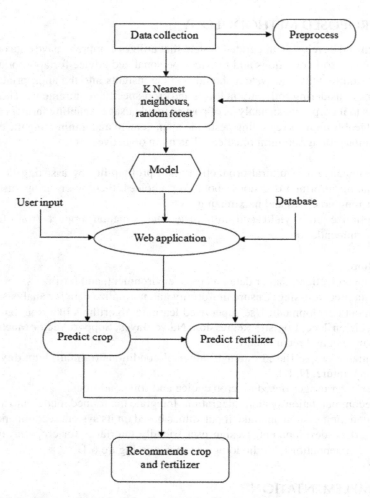

FIGURE 6.1   Proposed workflow for Crop and fertilizer recommendation system.

	N	P	K	temperature	humidity	ph	rainfall	pesticide	label
0	90	42	43	20.879744	82.002744	6.502985	202.935536	1	rice
1	85	58	41	21.770462	80.319644	7.038096	226.655537	1	rice
2	60	55	44	23.004459	82.320763	7.840207	263.964248	1	rice
3	74	35	40	26.491096	80.158363	6.980401	242.864034	1	rice
4	78	42	42	20.130175	81.604873	7.628473	262.717340	1	rice

FIGURE 6.2   Dataset.

like temperature, rainfall, pH, humidity, pesticide (previously utilized), N (nitrogen), P (phosphorous), K (potassium), label (crop name), which are crucial for understanding the nutrients in the soil and the meteorological conditions that support crop growth.

6.4.2 MACHINE LEARNING ALGORITHMS

1) *Random Forest (RF):* RF is a powerful ensemble learning method in machine learning that can be applied to regression and classification issues. It builds a huge number of decision trees during training by splitting while utilizing a random subset of the training set and a random selection of the characteristics. The forecasts from each of these produce the ultimate forecast, and more trees are added. In most cases, the model is used for classification and the mean for regression.

2) *K-Nearest Neighbor (KNN):* KNN is a straightforward and easily comprehensible supervised learning technique for regression and classification. Upon predicting a new data point via KNN, the algorithm locates the K training samples (neighbors) that are closest to the new point using a predetermined distance metric, most frequently the Euclidean distance. In a classification task, a newly added data point's class is predicted by the majority class among its KNNs. The average of the K closest neighbors' output values is typically the desired outcome of regression operations. K, or the number of neighbors, is one of the most significant variables influencing the model's effectiveness.

6.5 RESULTS

The accuracy of two distinct supervised machine learning algorithms, RF and KNN, are displayed in Figure 6.3, with corresponding accuracy percentages of 99.0% and 97.0%. The crop prediction based on user input parameters given by the RF machine learning algorithm is depicted in Table 6.2. The data set exploration can also be visualized in the graphs in Figures 6. 4 and 6.5.

With data-driven intelligence and advanced technology, this system stands to revolutionize farming practices, enabling farmers to make informed decisions that

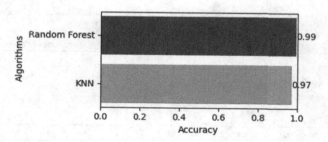

FIGURE 6.3 Accuracy comparison.

TABLE 6.2
Prediction Using Random Forest

N (Nitrogen)	P (Phosphorous)	K (Potassium)	Temperature	Humidity	H	Rainfall	Pesticide	Random Forest Prediction
27	21	40	19	80	2	95	1	Pomegranate
26	18	12	25	65	5	158	0	Pigeonpeas
25	16	10	24	90	6	168	1	Orange
120	22	31	25	80	6	164	1	Coffee
36	120	25	18	75	4	76	1	Grapes

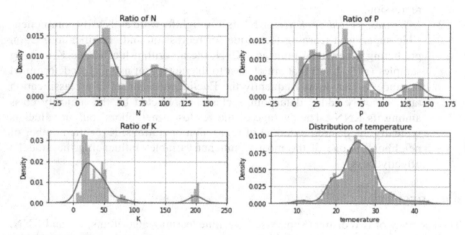

FIGURE 6.4   Dataset exploration.

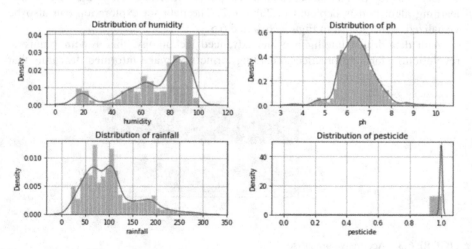

FIGURE 6.5   Dataset exploration.

not only enhance their yields and income but also contribute to the preservation of the ecosystem. Further, the proposed system can be extended to provide fertilization suggestions to farmers based on their usage.

From the above results we can observe that the RF-supervised learning algorithm shows the highest accuracy when compared with the KNN algorithm and also accurate crop prediction with the given parameters. Thus, it is proved that the proposed systems helps in enhancing the crop productivity.

6.6 CONCLUSION

The integrated crop and fertilizer recommendation system harnesses an extensive array of data on environmental conditions, soil attributes, and historical pesticide usage to empower farmers with tailored insights. By utilizing sophisticated analytics and machine learning, this system provides personalized recommendations for crop choices and eco-conscious fertilizer utilization. Its overarching objective is to maximize agricultural productivity and profitability while fostering sustainable farming practices that are mindful of environmental conservation.

REFERENCES

[1] C. Rajesh, V. H. S. Praveen, and G. S. Deepak, "Smart crop recommendation system using machine learning," *IOSR Journal of Electrical and Electronics Engineering*, vol. 18, no. 3, pp. 49–55, 2023.

[2] A. Motwani, P. Patil, V. Nagaria, S. Verma, and S. Ghane, "Soil analysis and crop recommendation using machine learning," in *2022 International Conference for Advancement in Technology (ICONAT)*, Goa, India, pp. 1–7, 2022.

[3] A. Jadhav, N. Riswadkar, P. Jadhav, and Y. Gogawale, "Crop recommendation system using machine learning algorithms," *International Research Journal of Engineering and Technology (IRJET)*, vol. 09, no. 04, pp. 1227–1230, 2022.

[4] C. Sagana, M. Sangeetha, S. Savitha, K. Devendran, T. Kavin, K. Kavinsri, and P. Mithun, "Machine learning-based crop recommendations for precision farming to maximize crop yields," in *2023 International Conference on Computer Communication and Informatics (ICCCI)*, IEEE, pp. 1–5, 2023.

[5] B. Gayathiri, P. Brindha, I. Karthika, E. Saranya, G. Rajeshkumar, and P. Rajesh Kanna, "Machine learning based crop suitability prediction and fertiliser recommendation system," in *2023 4th International Conference on Electronics and Sustainable Communication Systems (ICESC)*, pp. 1023–1028, 2023.

[6] R. Jeevaganesh, D. Harish, and B. Priya, "A machine learning-based approach for crop yield prediction and fertilizer recommendation," in *2022 6th International Conference on Trends in Electronics and Informatics (ICOEI)*, pp. 1330–1334, 2022.

[7] D. Israni, K. Masalia, T. Khasgiwal, M. Tolani, and M. Edinburgh, "Crop-yield prediction and crop recommendation system," *Available at SSRN 4111856*, 2022.

[8] D. A. Bondre and S. Mahagaonkar, "Prediction of crop yield and fertilizer recommendation using machine learning algorithms," *International Journal of Engineering Applied Sciences and Technology*, vol. 4, no. 5, pp. 371–376, 2019.

[9] S. R. Gopi and M. Karthikeyan, "Effectiveness of Crop Recommendation and Yield Prediction using Hybrid Moth Flame Optimization with Machine Learning", *Engineering Technology and Applied Science Research*, vol. 13, no. 4, pp. 11360–11365, Aug. 2023.

[10] S. M. Pande, P. K. Ramesh, A. Anmol, B. R. Aishwarya, K. Rohilla and K. Shaurya, "Crop Recommender System Using Machine Learning Approach," in 2021 5th International Conference on Computing Methodologies and Communication (ICCMC), Erode, India, pp. 1066–1071, 2021, doi: 10.1109/ICCMC51019.2021.9418351.

7 Empowering Agriculture
Harnessing the Potential of AI-Driven Virtual Tutors for Farmer Education and Investment Strategies

Soumya Priyadarshini Mishra and Munmun Mohanty

7.1 INTRODUCTION

With the introduction of artificial intelligence (AI) into conventional farming methods, the agricultural sector is undergoing a revolutionary wave in a period characterized by technological developments. The creation of virtual tutors designed to instruct farmers on how to make investments is one of the most important uses of artificial intelligence in agriculture. This article explores how AI-driven virtual tutors are changing the learning experience for farmers and enabling rural populations to make well-informed investment decisions. It also dives into the exciting world of these tutors. It is clear as we work our way through the complexities of this cutting-edge strategy that combining AI with agriculture has enormous promise to improve global farmer output, long-term viability, and financial results.

This chapter sets out to investigate the revolutionary possibilities AI has for enhancing agriculture. By concentrating on the incorporation of virtual instructors, we explore how these smart technologies might act as a change agent by providing farmers with tailored and flexible learning opportunities. Additionally, we look at how AI can optimize approaches to investing and give users statistical information to make better decisions. Our goal in doing this research is to demonstrate the possibility of AI-powered virtual instructors as effective instruments that may support farmers, encourage environmentally friendly methods, and open the door to a resilient and productive farming destiny. It is essential to incorporate innovation into agriculture in order to create an ecologically sound and prosperous industry.

A wide spectrum of AI-driven applications in high-tech agriculture can be used, including post-harvest procedures, distribution logistics, planting, harvesting, crop monitoring, and seed selection [1]. Artificial Intelligence has the capability to identify irregularities and deficiencies in soil nutrients and may offer farmers instantaneous guidance on the appropriate timing of pesticide application, fertilizer application, and irrigation. Farmers may increase productivity, efficiency, and

DOI: 10.1201/9781003484608-7

sustainability in their operations by automating processes, gathering real-time data, and making well-informed decisions by connecting physical devices and sensors to the internet [2]. As technology advances, businesses that are working to improve AI and machine learning-based goods and services—such as training data for drones, automated machinery, and agriculture—will be able to offer this industry more beneficial uses. Assisting the globe in addressing the challenges of food production for the expanding population will be a huge help [3]. AI has the potential to revolutionize agricultural production and enhance efficiency, sustainability, and productivity in several ways [4]. Climate change is affecting agricultural systems, such as modified precipitation patterns, a rise in the frequency of extreme weather events, and variations in the temperature ranges. These modifications cause crop cycle disturbances, lower yields, and increased farmer vulnerability [5]. One potential solution to the problems caused by insufficient labor in rural regions is smart farming. Due to the population concentration in cities, farming may now be done remotely. Farmers are able to maximize conditions for crop yield and quality by employing big data to do crop modeling and operations [6]. Assuring farmers of a fair return and giving them a sense of security and confidence are the main goals of agricultural price policies like Minimum Support Price (MSP). Nevertheless, India's agricultural price policy has achieved this goal to some extent, but it has also had some negative repercussions and added to the inflationary trends in the economy [7]. Developments in digital agriculture management emphasize how AI, IoT devices, machine learning, and sensor technologies are all integrated to improve farming production and efficiency. There is a focus on the usage of robotics, smart irrigation systems that employ real-time data to optimize agricultural operations, and drones for crop monitoring. High-speed 5G networks, in particular, affect connectivity and data transfer in rural areas, making these smart farming solutions possible [8]. Although AI technology has a lot of potential to improve agriculture sustainability through data-driven decision-making, supply chain optimization, weather forecasting, precision farming, labor efficiency, and crop monitoring, it also has drawbacks. In order to guarantee equal access to AI solutions, it is imperative to bridge the digital divide and overcome financial restrictions in areas like the global south that have restricted access to technology [9]. Supply chain efficiency, precision agriculture, and crop management are all being transformed by AI and machine learning (ML) algorithms. Making use of information from multiple sources, such as satellites, drones, and sensors, these technologies allow for improved crop yields, more efficient use of resources, and better livestock health monitoring [10]. Fixed investments and liquid investments both benefit from financial literacy. Due to variations in the size of their farms and the insurance they choose, farmers have differing benefits from financial literacy on liquid and fixed investments. The impacts of financial literacy on fixed and liquid investments are mediated by risk preferences [11]. More visual data is now available from more sources, including satellites, drones, and ground-based imaging tools. These data are crucial for understanding crop and soil health as well as animal monitoring. In order to support precision agriculture and lower the usage of chemicals, AI systems can analyze this vast amount of data and identify diseases, pests, and nutritional deficiencies early on [12]. Input subsidies continue to boost agricultural productivity more than public investments do. Additionally, it has been discovered that the most effective input subsidy is power, followed by fertilizer. Government spending on input subsidies is therefore justified since it guarantees that all farmers will have access to reasonably priced agricultural inputs. Ample investment in agricultural infrastructure, along with targeted subsidies, may result in long-term agricultural development in India [13]. AI is a significant area of contemporary computer science that is becoming more and more vital to the advancement of contemporary agriculture. The fields of application for intelligent robots based on AI technology are expanding, and the research and development of these machines are maturing [14]. The arrival of self-driving machines explains how AI-powered solutions improve efficiency and output. Artificial intelligence (AI) maintains effective water and energy management with a focus on environmental sustainability and resource efficiency. AI helps farmers by distributing knowledge, filling in information gaps, and encouraging well-informed decision-making [15]. In precision agriculture, AI and robots are highlighted for their contributions to data-driven decision-making, automated farming operations, and sustainable farming methods. The cooperation of robotic systems, sophisticated sensors, and AI algorithms platforms facilitate prompt data gathering, analysis, and focused actions, which promote effective resource management and increased crop yield [16]. AI is still in its early stages of development. Modern stand-alone and hybrid algorithms are being created and used to solve real-world issues pertaining to agriculture. Similarly, these models' efficiency and modeling accuracy improve with each new development, and their parametric requirements also drop [17–19].

7.2 FARMERS SERVICES MODELS

Chatbots: At the moment, the retail, travel, media, and insurance industries utilize chatbots, or virtual assistants, driven by artificial intelligence. However, agribusiness could also benefit from this advancement in technology by offering advice and solutions to farmers on particular issues. Through interactive voice chat in their local tongues, this program enables farmers to get the answers to their questions. For ongoing and context-sensitive learning, the chatbot engine is powered by both supervised and reinforced machine learning algorithms. Thus, the chatbot responds to the majority of general inquiries before allowing a human operator to assist with any inquiries that are specific to the user.

Agri-E-Calculator: This clever tool assists the savvy farmer in selecting the most economical crop by taking into account a number of dependent variables. Using the clever calculator, the farmer only needs to select the crop he wants to grow throughout the area of his farm that he wants covered. The e-calculator then detects and takes all other necessary inputs according to different interdependence factors and outputs the projected results. The resulting outcome offers helpful information on the expense and volume of fertilizer products, water, seeds, cultivation supplies, and labor allocation on a calendar graph of the crop life cycle. It also includes crop yield estimates and projected market prices at the conclusion of harvest. The producer's record-keeping system and the previously mentioned outside sources of data provide all necessary inputs, which might be either linear or non-linear in character. Machine learning algorithms are used to process the inputs and produce a projection together with a viability assessment, allowing farmers to select the crop they want to cultivate.

Service for Crop Care: The advice for agricultural care services begins at the point of planting seeds and ends after reaping is completed. Intelligent AI algorithms are used to analyze the complicated data structures collected from IoT devices in the field, combined with data acquired from websites and domain professional inputs where necessary. The general correction item is calculated using PID (Proportional Integral & Differential) following the examination of all available data.

Price Forecasting and Market Advice: These services lessen the chance of price loss and protect farmers from market fluctuations. Throughout the whole crop lifecycle, farmers receive forecast pricing and demand statistics based on statistical data gathered from multiple sources. As a result, farmers are able to make better decisions about when to release their goods onto the market.

Crop Loan and Insurance Service: This program assists farmers in determining the viability of obtaining an agricultural loan, as well as the eligibility requirements, loan limitation, as well as processing assistance, based on a careful estimate generated for the proposed crop. Obtaining crop insurance is also beneficial as a preventive measure against unsuccessful crops brought on by unpredictability or disasters.

7.3 AI TECHNOLOGIES APPLIED IN FARMING CURRENTLY

Blue River Technology: Established in 2011, this California-based business builds contemporary agricultural technology that uses fewer chemicals while costing less money by combining robots, machine vision, and AI. Robotics allows intelligent robots to act, while computer vision recognizes and determines the specific needs of each unique plant. The application of appropriate herbicides inside the proper buffer surrounding the plant, as well as the usage of sensors that identify the kind of invasive plants. When photographs are taken, and the machines can be trained in various weeds, the cameras and sensors employ algorithmic learning. Furthermore, the appropriate herbicides are applied accurately in accordance with the encroaching region. A robot named See & Spray was created by Blue River Technology, and it is said to use artificial intelligence to track and accurately eradicate weeds on cotton crops. Resistance to herbicides may be avoided with the aid of precise application.

FarmBot: Founded in 2011; by providing environmentally aware consumers with the tools to cultivate crops at home, this startup has elevated precision farming to a new level. At $4000, the device FarmBot assists the owner in managing the entire farming process independently. This physical bot uses an open-source software system to handle everything from seed planting to weed detection, soil testing, and plant watering.

Harvest CROO Robotics – Crop Harvesting: A robot has been created by Harvest CROO Robotics to assist strawberry farmers with crop picking and packing. Key farming regions like California and Arizona have reportedly lost millions of dollars in revenue due to a lack of labor. By picking up strawberries, the robot helps farmers cut down on the expense of harvest labor. Because strawberries must be harvested within a specific time frame, skilled workers are required. Harvests CROO Robotics is confident that their creation will enhance quality, lower energy costs, boost harvests, and save funds.

Plantix: An agricultural tech business based in Berlin, PEAT created an app called Plantix to diagnose plant diseases. It helps identify probable faults and nutrient deficiencies in soil. The application uses photos to identify plant diseases. First, a smartphone takes a picture, which is compared to a server image to determine the plant's health. In order to treat plant diseases, the program makes use of AI and machine learning.

Prospera: Formed in 2014. The farming industry has been completely transformed by this Israeli firm. It has created a cloud-based solution that compiles all of the data that farmers now possess, including aerial photos, soil and water sensors, and more. After that, it integrates it with an in-field tool to interpret everything. The Prospera gadget, which is driven by a range of sensors and technology, including computer vision, may be utilized in fields and greenhouses. These sensors' inputs are utilized to create predictions by determining a correlation between various data labels.

7.4 AI AND ITS PRIORITY FOR THE INDIAN GOVERNMENT IN RURAL AGRICULTURE

Agriculture Secretary Sanjay Agarwal said that AI and big data will play significant roles in the agriculture industry in the upcoming years since data is "key to targeted development" during his speech at the third "India Agricultural Outlook Forum 2019" with the topic "Universal Basic Income for Farmers." Through the registration procedure for initiatives at the national level, the Indian government is gathering vast amounts of data about farmers. Digitization of land holding details is also underway. To learn more about the state of farmers and their crops, the government is looking into data from the Kisan Credit Card (KCC), soil health cards, and farm insurance. AI, modeling tools, and remote sensing imagery are being used by the Pradhan Mantri Fasal Bima Yojana (PMFBY) to speed up the settlement of claims. Nearly 20 million farmers are covered by PMFBY.

Although it is difficult to view a farmer as a businessperson, that is exactly what he is. He manages a company that needs finance, strategy, and direction. Agriculture-related venture financing has always existed; government programs and different banks provide credit lines to farmers all around the nation. Additionally, farmers can apply for a number of incentives and subsidies. In an effort to strengthen the rural economy, the National Bank for Agriculture and Rural Development (NABARD) started the practice of offering farmers financial options and loan facilities back in the 1980s. One of NABARD's most prominent programs is the Kisan Credit Card, where the amount of the loan is determined by a number of variables, including cultivation costs, yield percentages, maintenance costs, and more. The Agriculture Infrastructure Fund, a pan-India Central Sector Scheme recently approved by the central government, will offer a medium- to long-term debt financing facility through interest subvention and financial support for investments in feasible projects for post-harvest management infrastructure and community farming assets.

Nevertheless, agri-financing has numerous difficulties. Banks are hesitant to lend to farmers due to concerns about Non-Performing Assets (NPAs), which are mostly caused by loan defaults. Not all of this distrust is unjustified. One of the main causes is the unreliability of farmers; for every agri-lender, the key question is: "Will the loaner repay the money with interest on time? Will the loaner's farming endeavors continue to be profitable in the future, allowing me to confidently offer another line of credit?"

A lender's replies to the aforementioned queries typically do not inspire confidence in them. And these are only a few issues and difficulties with agricultural funding. However, there remains a financial shortfall for the industry. Is technology able to offer an answer?

In Indian agriculture, this is a new trend. Many business owners are investigating the ways in which AI might be applied to evaluate farms more accurately and provide the best possible credit line. CropIn, a Bangalore-based company, is doing just that. SmartRisk, a tool developed by CropIn using AI and machine learning, recognizes cropping trends and projects crop futures. Utilizing crop stage identification, examine the pattern of sowing and harvesting as well. CropIn's smart-risk technology assists in determining a farmer's creditworthiness prior to loaning money and in monitoring them afterward. Financial firms can assess a farmer's quality of service, business practices, and risk management by carefully examining the data points. Agricultural health and water stress are important markers of agricultural performance that offer information for risk assessment in a given area. Additionally, SmartRisk assists the company in obtaining these data points. Bank workers geotag and audit necessary farm plots, while CropIn provides clients with historical reports on crop health, yield, and water stress and continues to monitor the crop.

Banks can use cadastral maps to determine the best plots within a 50-kilometer radius of a branch by performing health analyses on the plots and making them available for examination. It is feasible to process remote sensing-based images over the whole project area to display the health, estimated yield, and crops that have been grown. The entire project area is subjected to remote sensing-based image processing for post-code level analysis, which displays the health and an estimated yield with alerts and periodic checks. These indicators aid in lowering operating costs and streamlining the loan disbursement process. Reducing the number of personnel needed for field audits results in faster loan collection and efficient NPA management. Financial institutions, such as banks, Non-Banking Financial Companies (NBFCs), and Micro Finance Companies (MFIs), can enhance the loan qualification and disbursement processes, anticipate NPAs, monitor risk in real-time through crop growth analysis, and confidently expand their lending portfolios to new geographic areas. Banks, NBFCs, and microfinance institutions can better understand agricultural credit risk by region, optimize loan disbursement, manage loan delinquency, manage NPAs, and manage loan collection with the aid of customized, targeted data-driven solutions. This data can also be used for loan underwriting, and drought and water stress can be assessed more thoroughly.

The World Economic Forum's Artificial Intelligence for Agriculture Innovation (AI4AI) program is addressing these issues by promoting the application of AI and related technologies for agricultural developments, hence aiding India's agricultural revolution. This program, spearheaded by the Center for the Fourth Industrial Revolution (C4IR) India, unites government, academic, and commercial officials to devise and execute creative ideas for the agriculture industry.

The 'Saagu Baagu' pilot, which was designed in collaboration with the Telangana state government and is being carried out by Digital Green in the Khammam district, is among the most effective applications of the AI4AI initiative. It has received support from the Bill and Melinda Gates Foundation. For over 7,000 farmers, the project has significantly enhanced the value chain for chilies. Telangana's state government, which established the nation's first framework for agri-data management and exchange as well as other supportive policies, has been instrumental in this shift. Saagu Baagu has shown impressive outcomes during its initial period of operation. A 21% increase in chili yields per acre, a 9% decrease in pesticide use, a 5% decrease in fertilizer use, and an 8% increase in unit prices as a result of quality improvements were observed by farmers involved in the program. Farmers' revenues have increased by more than INR 66,000 (about 800 USD) per acre per crop cycle as a result of these changes, nearly doubling their income. These numbers demonstrate Saagu Baagu's efficacy as well as its role in productive and environmentally friendly farming practices. Based on these achievements, the Telangana government broadened the scope of Saagu Baagu in October 2023. Presently, the project intends to affect 500,000 farmers in ten districts, covering five distinct crops. This growth is a calculated effort to optimize the advantages of cutting-edge farming innovations, which might completely change the region's agricultural environment.

7.5   USING AI TO IMPROVE INVESTMENT STRATEGIES

AI is shown itself to be a revolutionary force in the agriculture industry when it comes to optimizing investment plans, in addition to its function in farmer education. This section examines the ways in which AI-powered technologies are transforming financial modeling, risk management, and decision-making procedures and offering stakeholders insightful information for wise and strategic investment choices.

Predictive Analytics for Market Trends: Artificial Intelligence is excellent at analyzing large volumes of data and spotting trends that conventional analysis might miss. This capacity is used in the field of agriculture to estimate market trends using predictive analytics. AI-powered virtual tutors examine global demand-supply dynamics, historical market data, and weather trends. This helps investors and farmers to plan ahead for changes in the market, act quickly, and match their capital to new ventures.

Risk Mitigation Methods: Robust risk management methods are required due to the inherent hazards in agriculture, which include weather uncertainty, commodity price volatility, and geopolitical issues. AI-powered virtual tutors assist by evaluating risk factors, spotting possible dangers, and suggesting countermeasures. These tools help investors and farmers create plans that are strong enough to face unforeseen obstacles by using advanced risk models and historical data.

Financial Modeling and Resource Optimization: Complex financial factors are frequently taken into account when making investment decisions in agriculture. AI improves financial modeling by adding a variety of variables, including labor costs, input costs, and market pricing. Virtual tutors offer cost-effective techniques that improve overall profitability and help optimize resource allocation. By ensuring that investments are in line with stakeholders' financial goals, this data-driven strategy maximizes returns on investment.

Investment in Precision Agriculture: The AI-powered concept of precision agriculture demands efficient and targeted resource management. Investors can learn and use precision agriculture techniques with the aid of AI-powered virtual tutors. These tools facilitate the deployment of technologies like drones, sensors, and automated machinery, maximizing the return on agricultural investments. Additionally, these technologies support resource-efficient and sustainable farming practices.

Diversification Strategies: AI makes it easier to analyze a variety of sources of information, which empowers investors to decide how to diversify their portfolios in an informed manner. Virtual instructors evaluate variables, including market demands, soil health, and climate, to provide guidance on animal selection and crop diversification. By reducing the risks connected with relying too heavily on a particular crop or commodity, this strategy encourages a more robust and diverse investment portfolio.

Decision Support and Real-Time Monitoring: Because agriculture is a dynamic industry, it requires real-time decision support and monitoring. AI-powered virtual tutors facilitate quick updates on important metrics to stakeholders, enabling them to make decisions quickly. These technologies offer useful information that can be used to improve the flexibility of investment plans, such as modifying irrigation schedules in response to weather forecasts or abrupt changes in the market.

7.6   FUTURE SCOPE

Looking ahead, the application of AI-powered virtual tutors in agriculture has great potential to usher in a new era of creativity, sustainability, and adaptability. Future prospects appear promising due to the continuous progress in artificial intelligence and the growing utilization of digital technology in the agricultural sector.

Complex Algorithms for Learning: The advancement and complexity of learning algorithms will determine the direction of AI-powered virtual tutors in the future. Virtual instructors' ability to comprehend the complex requirements of individual farmers will advance as machine learning and deep learning models develop further. As more sophisticated algorithms are developed, these tutors will be able to offer even more context-specific and individualized advice, meeting the particular difficulties that farmers in various areas experience.

Integration of Emerging Technologies: The agricultural environment is about to change as AI and other emerging technologies combine. Virtual instructors' skills will be improved by integration with edge computing, blockchain, and the IoT. The seamless integration of real-time data from sensors, drones, and smart devices would enable more precise decision-making and prompt reaction to shifting farm conditions.

Democratization of AI: In the future, AI-powered virtual tutors should be widely available to farmers across all areas and scales, providing them with revolutionary tools. Initiatives aimed at closing the digital gap, enhancing internet availability, and advancing reasonably priced technologies will enable even the most isolated and neglected farming communities. This diversity will help create equity and level the playing field.

Autonomous Agricultural Systems: The development of autonomous agricultural systems may follow the emergence of virtual tutors. Daily farming operations could be revolutionized by combining robotic technologies, autonomous machines, and AI-driven virtual teachers. From planting and harvesting to irrigation and pest management, self-governing systems led by virtual tutors have the potential to improve productivity, lower labor costs, and maximize resource use.

Improved Predictive Analytics: Developments in this field will improve virtual tutors' capacity to predict crop yields, weather patterns, and market trends. More exact and detailed forecasts in the future should enable investors and farmers to act quickly on well-informed information. This increased capacity for prediction will help agriculture become more resilient overall and manage risk better.

Collaborative Decision Platforms: These platforms, where virtual tutors assist in communication and knowledge exchange between farmers, academics, and stakeholders, are expected to become more prevalent in the agricultural scene in the future. These platforms have the potential to cultivate an international community of agricultural practitioners, facilitating the sharing of creative solutions, best practices, and up-to-date information, ultimately advancing the cause of global food security.

7.7   CONCLUSION

At the ever-changing intersection of farming and AI, this chapter has explored the transformative potential of AI-driven virtual tutors to improve farmer education and investment strategies. Science and agriculture are working together to usher in a new era of sustainable economic growth and production of food. The field is being redefined by data-driven insights and customized learning experiences. By offering individualized, tailored learning experiences, AI-driven virtual tutors are revolutionizing the field of farmer education. These virtual companions play a vital role in providing farmers with the information and abilities required to successfully traverse the intricacies of contemporary agriculture, including anything from advice on precision farming to the promotion of sustainable practices. By keeping farmers up to date on innovations, the continuous learning model fosters adaptation in the face of changing obstacles. AI is a key component in improving the modeling of finances, handling risks, and processes for making decisions in the financial investment space. Strong methods for risk mitigation help stakeholders remain resilient in the face of uncertainty, while predictive analytics helps them predict market trends. Allocating resources optimally and advancing precision agriculture tactics guarantee that investments are in line with both budgetary goals and environmentally friendly methods. This chapter has emphasized the practical applications of AI-driven virtual tutors in agriculture through examples. The revolutionary potential of these technologies is clear across many agricultural contexts, from large-scale investors making strategic decisions backed by predictive analytics to smallholder farmers in distant places having access to essential knowledge.

REFERENCES

1. Vashishth, T. K., Sharma, V., & Kumar, B. (2024). Artificial Intelligence (AI)-Integrated Biosensors and Bioelectronics for Agriculture. In A. Khang (Ed.), Agriculture and Aquaculture Applications of Biosensors and Bioelectronics (pp. 158–183). IGI Global. https://doi.org/10.4018/979-8-3693-2069-3.ch008.

2. Suma, K. G., Aswini, J., Sunitha, G., & Balaji, K. (2023). AI-Driven Applications in High-Tech Agriculture. In A. Khang (Ed.), Handbook of Research on AI-Equipped IoT Applications in High-Tech Agriculture (pp. 23–37). IGI Global. https://doi.org/10.4018/978-1-6684-9231-4.ch002.

3. Huchchannanavar, S., & Anuja, D. (2023, Oct). Chapter 3 Artificial Intelligence (AI) and IoTs in Agriculture: A Concept and Reality. Emerging trends in Agronomy (pp. 39–52). Biotech Books.

4. Sharadkumar Rudrawar, S. (n.d.). AI for Everyone: Applications. https://doi.org/10.5281/zenodo.8287514

5. Pisal, D. S., Patil, S., Mirji, H., & Phalke, V. S. (n.d.). AI-Powered Innovations in Agriculture Transforming Farming Practices and Yield Optimization. Indian Journal of Technical Education, 46(126). www.researchgate.net/publication/376046268

6. Amruddin, A., Mahmood, M. R., Supardjo, D., Ibrahim, A., & Susilatun, H. R. (2024). Analysis of Climate Change Impacts on Agricultural Production and Adaptation Strategies for Farmers: Agricultural Policy Perspectives. Global International Journal of Innovative Research, 2(1). https://doi.org/10.59613/global.v2i1.50

7. Yadav, A. (n.d.). Unleashing the Potential of Agriculture: AI-Powered Smart Farming. www.researchgate.net/publication/372768753

8. Araghyadeep Das, N., & Madhav, P. S. M. (2020). Doubling Farmers' Incomes: Issues, Strategies and Recommendations. In Doubling Farmers' Incomes: Issues, Strategies and Recommendations. In What is Minimum Support Price and does it Really Benefit Farmers? (pp. 1–14). AkiNik Publications. https://doi.org/10.22271/ed.book.938

9. Issa, A. A., Majed, S., Ameer, S. A., & Al-Jawahry, H. M. (2024). Farming in the Digital Age: Smart Agriculture with AI and IoT. E3S Web of Conferences, 477. https://doi.org/10.1051/e3sconf/202447700081

10. Sakapaji, S. C., & Puthenkalam, J. J. (2023). Harnessing AI for Climate-Resilient Agriculture: Opportunities and Challenges. European Journal of Theoretical and Applied Sciences, 1(6), 1144–1158. https://doi.org/10.59324/ejtas.2023.1(6).111

11. Hussein, A. H. A., Jabbar, K. A., Mohammed, A., & Jasim, L. (2024). Harvesting the Future: AI and IoT in Agriculture. E3S Web of Conferences, 477. https://doi.org/10.1051/e3sconf/202447700090

12. Liu, G., Li, Y., & Xu, D. (2024). How Does Financial Literacy Affect Farmers' Agricultural Investments? A Study from the Perspectives of Risk Preferences and Time Preferences. *Applied Economics*, 1–15. https://doi.org/10.1080/00036846.2024.2313596

13. Yang, W., Wang, J., Hu, C., Yang, M., & Fei, J. (2023). Research on Application of Intelligent Robot Based on AI Technology in Agriculture. *Highlights in Science, Engineering and Technology*, 76, 668–673. https://doi.org/10.54097/a6xva453.

14. Kumar, D., Rank, P. H., Patel, R., & Vekariya, P. B. (n.d.). The Growing Potential of Artificial Intelligence (AI) Based Technologies in Agriculture. www.researchgate.net/publication/377245205

15. Verma, R. K., & Kishor, K. (2024). Image Processing Applications in Agriculture with the Help of AI (pp. 162–181). https://doi.org/10.4018/979-8-3693-0782-3.ch010

16. Zafar, S., Aarif, M., & Tarique, Md. (2023). Input Subsidies, Public Investments and Agricultural Productivity in India. *Future Business Journal*, 9(1). https://doi.org/10.1186/s43093-023-00232-1

17. Anshuman, J. (n.d.). Precision Agriculture and the Role of AI and Robotics. www.researchgate.net/publication/378262405

18. Hammad, M., Shoaib, M., Salahudin, H., Baig, M. A. I., & Ali, M. U. (2023). Use of AI for Disaster Risk Reduction in Agriculture (pp. 461–488). https://doi.org/10.1007/978-981-99-1763-1_22

19. IGI Global. AI-Driven Applications in High-Tech Agriculture (pp. 23–37). https://doi.org/10.4018/978-1-6684-9231-4.ch002

8 Enhancing Agricultural Ecosystem Surveillance through Autonomous Sensor Networks

P. Venkata Kishore, K. Sree Latha,
D. Naveen Kumar, M. Dilip Kumar,
S. Vijay Kumar, and K.V.B. Reddy

8.1 INTRODUCTION

This paper designs a low-cost agriculture ecosystem for irrigation to reduce the manual monitoring of the field using sensors. The data is also displayed on a Liquid-Crystal Display (LCD) [1]. This paper makes use of a Peripheral Interface Controller (PIC) microcontroller, the main controlling device like a soil moisture sensor, which is used to detect the moisture content in the soil. A water level sensor is used to detect the water level in the tank. An LM5 Temperature sensor is used to detect the temperature of the motor. If the motor works without water, the motor will get heat. The controlling device of the whole system is the PIC microcontroller [2]. Whenever the sensor unit gets input from respected sensors like temperature sensors, water level indicators, and soil moisture sensors, these inputs are fed to the microcontroller [3]. The microcontroller performs appropriate tasks related to the data received, like motor ON/OFF control. If soil moisture is low, or water level is low, this system will switch on the water motor. If the water level is at an appropriate level, then the motor is OFF automatically [3,4]. If the temperature of the motor crosses the set limit, then the microcontroller will switch OFF the motor automatically. The appropriate values are displayed on LCD.

8.2 HARDWARE DESCRIPTION

When making a PCB, you have the option of making a single-sided board or a double-sided board. Single-sided boards are cheaper to produce and easier to etch but much harder to design for large devices. If a lot of parts are being used in a small space, it may be difficult to make a single-sided board without umpiring over traces with a cable. While there's technically nothing wrong with this, it should be avoided if the signal traveling over the traces is sensitive (e.g., audio signals). A double-sided board is more expensive

DOI: 10.1201/9781003484608-8

to produce professionally and more difficult to etch on a DIY board, but it makes the layout of components a lot smaller and easier. It should be noted that if a trace is running on the top layer, check with the components to make sure you can get to its pins with a soldering iron. Large capacitors, relays, and similar parts that do not have axial leads can NOT have traces on top unless boards are plated professionally. Ground-plane or other special purposes for one side [5]. When using a double-sided board, you must consider which traces should be on what side of the board. Generally, put power traces on the top of the board, jumping only to the bottom if a part cannot be soldiered onto the top plane (like a relay), and vice- versa [6]. Some tasks like power supplies or amps can benefit from having a solid plane to use for the ground. In power supplies, this can reduce noise, and in amps, it minimizes the distance between parts and their ground connections and keeps the ground signal as simple as possible. However, care must be taken with stubborn chips such as the TPA6120 amplifier from TI. The TPA6120 datasheet specifies not to run a ground plane under the pins or signal traces of this chip as the capacitance generated could affect performance negatively. PIC compiler is software used where the machine language code is written and compiled. After compilation, the machine source code is converted into hex code which is to be dumped into the microcontroller for further processing. PIC compiler also supports C language code. It is important that you know C language for microcontrollers, which is commonly known as Embedded C. As we are going to use a PIC compiler, hence we also call it PIC C. The PCB, PCM, and PCH are separate compilers. PCB is for 12-bit opcodes, PCM is for 14-bit opcodes, and PCH is for 16-bit opcode PIC microcontrollers [7]. Due to many similarities, all three compilers are covered in this reference manual. Features and limitations that apply to only specific microcontrollers are indicated within. These compilers are specifically designed to meet the unique needs of the PIC microcontroller. This allows developers to quickly design applications software in a more readable, high-level language. When compared to a more traditional C compiler, PCB, PCM, and PCH have some limitations. As an example of the limitations, function recursion is not allowed. This is due to the fact that the PIC has no stack to push variables onto and also because of the way the compilers optimize the code [8]. The compilers can efficiently implement normal C constructs, input/output operations, and bit-twiddling operations. All normal C data types are supported along with pointers to constant arrays, fixed point decimals, and arrays of bits. PIC C is not much different from a normal C program. If you know assembly, writing a C program is not a crisis. In PIC, we will have a main function in which all your application-specific work will be defined. In the case of embedded C, you do not have any operating system running there. So, you have to make sure that your program or main file should never exit. This can be done with the help of a simple while (1) or for (;;) loop, as they are going to run infinitely [9]. We have to add a header file for the controller you are using. Otherwise, you will not be able to access registers related to peripherals.

8.3   COMPILATION AND SIMULATION

For the PIC microcontroller, the PIC C compiler is used for compilation. The compilation steps are as follows: Open PIC C compiler. You will be prompted to choose a name for the new task, so create a separate folder where all the files of your project will be stored, choose a name, and click save. Click task, New, and something the box named 'Text1' is where your code should be written later. Now you have to click 'file, save as' and choose a file name for your source code ending with the letter '.c'. You can name as 'tasks' for example and click save. Then you have to add this file to your task work. You can then start to write the source code in the window titled 'tasks'. Then before testing your source code, you have to compile your source code and correct eventual syntax errors. By clicking on the compile option.hex file is generated automatically. This is how we compile a program for checking errors, and hence, the compiled program is saved in the file where we initiated the program [10].

After compilation, the next step is simulation. Here, the first circuit is designed in Express PCB using Proteus 7 software, and then simulation takes place, followed by dumping. The simulation steps are as follows: Open Proteus 7 and click on IS1S6. Now it displays PCB, where the circuit is designed using a microcontroller [11]. To design circuit components are required. So, click on the component option. 10. Now click on the letter 'p', then under that, select PIC16F877A, other components related to the project, and click OK. The PIC 16F877A will be called your *"Target device"*, which is the final destination of your source code.

Dumping steps: The steps involved in dumping the program edited in protest 7 to microcontroller are shown below:

Figure 8.1 shows initially, before connecting the program dumper to the microcontroller kit, the window appears as shown below.

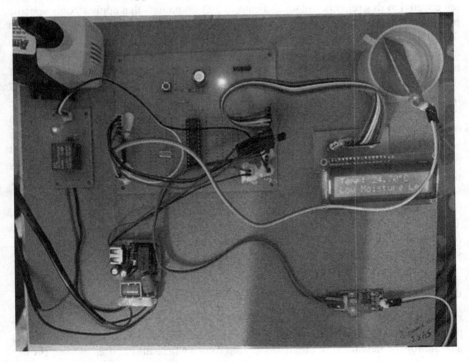

FIGURE 8.1   Picture of program dumper window.

Figure 8.2 shows the Select Tools option and click on Check Communication for establishing a connection as shown below.

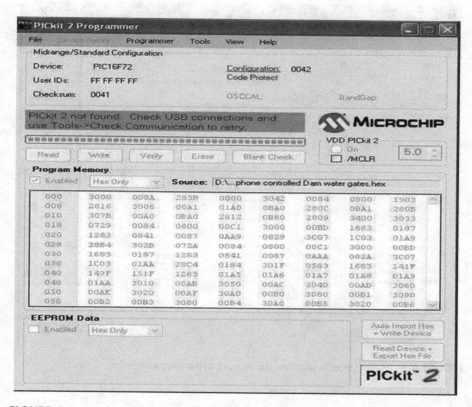

FIGURE 8.2 Picture of checking communications before dumping the program into the microcontroller.

Figure 8.3 shows, after connecting the dumper properly to the microcontroller kit the window appears as shown below.

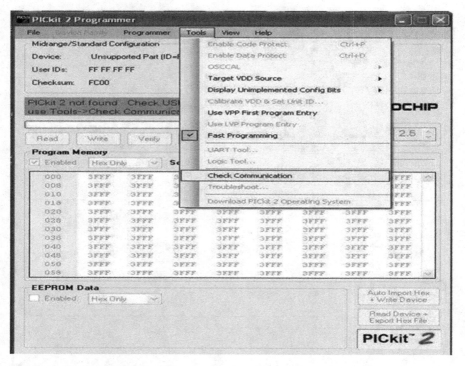

FIGURE 8.3   Picture after connecting the dumper to the microcontroller.

Figure 8.4 shows, again by selecting the Tools option and clicking on Check Communication the microcontroller gets recognized by the dumper and hence the window is as shown below.

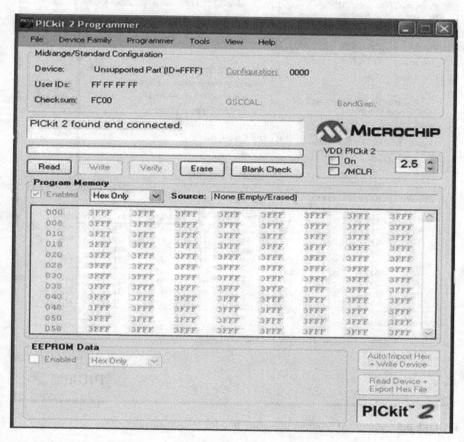

FIGURE 8.4 Picture of dumper recognition to the microcontroller.

Figure 8.5 shows import the program, which is '. hex' file from the saved location by selecting file option and licking on 'Import Hex' as shown in below window.

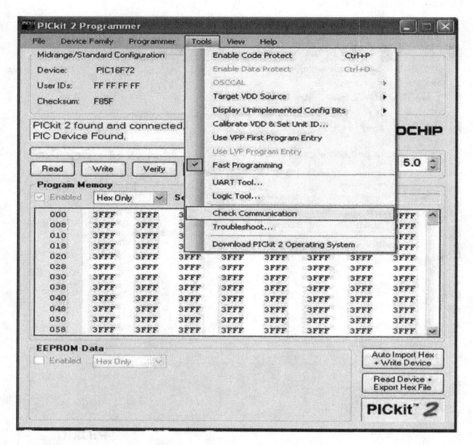

FIGURE 8.5   Picture of program importing into the microcontroller.

Figure 8.6 shows, after clicking on the 'Import Hex' option we need to browse the location of our program and click the 'prog.hex' and click on 'open' for dumping the program into the microcontroller. After the successful dumping of program the window is as shown below.

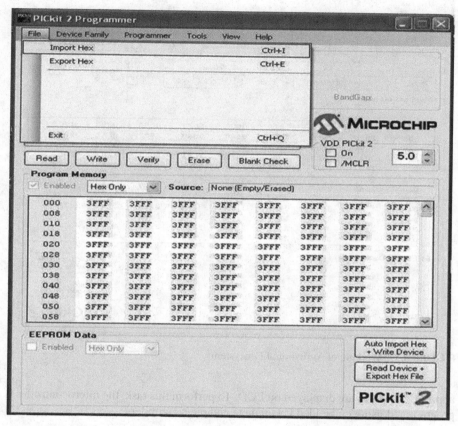

FIGURE 8.6   Picture after program dumped into the microcontroller.

8.4   RESULTS

The controlling device of the whole system is the PIC microcontroller. Whenever the sensor unit gets input from respected sensors like the temperature sensor, water level indicator and soil moisture sensor, these inputs are fed to the microcontroller. The microcontroller performs appropriate tasks related to the data received, like motor ON/OFF control, as shown in Figure 8.7. If soil moisture is low or water level is low this system will switch on the water motor. If the water level is at the appropriate level, then the motor is OFF automatically [12]. If the temperature of the motor crosses the set limit, then the microcontroller will switch OFF the motor automatically. The

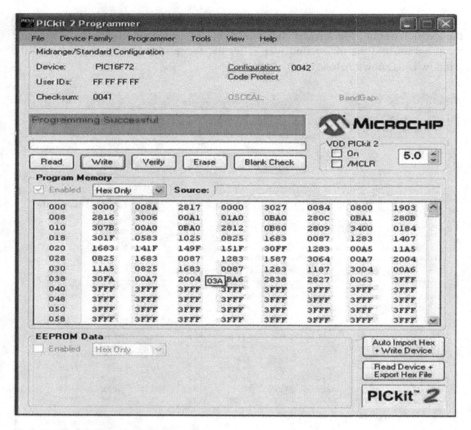

FIGURE 8.7   Working of Agricultural Ecosystem.

appropriate values are displayed on LCD. To perform this task, the microcontroller is programmed using embedded C language and pic c compilers.

8.5   CONCLUSION AND FUTURE SCOPE

8.5.1   CONCLUSION

This section showed a development to integrate the features of all the hardware components used. Every module's presence has been reasoned out and placed carefully, thus contributing to the unit's best workings. Secondly, the project has been successfully implemented using highly advanced integrated circuits (ICs) with the help of growing technology. Thus, the project has been successfully designed and tested.

8.5.2   FUTURE SCOPE

The project can be extended by adding GSM to get alert messages. We can add a Raspberry pi processor and some sensors like the NPK sensor and the PH sensor,

So the user can check his field every time and keep the field safe. The project can be extended by adding a SOLAR panel, a free sun source.

REFERENCES

[1] S. Cesco, P. Sambo, M. Borin, B. Basso, G. Orzes, F. Mazzetto. Smart agriculture and digital twins: Applications and challenges in a vision of sustainability. *European Journal of Agronomy*, 146 (2023).

[2] K.A. Mohamed Junaid, Y. Sukhi, S. Anita. Low-cost smart irrigation for agricultural land using IoT. *IETE Journal of Research*, 70(3), (2023). https://doi.org/10.1080/03772 063.2023.2178535.

[3] S.H. Muhie. Novel approaches and practices to sustainable agriculture. *Journal of Agriculture and Food Research*, 10 (2022).

[4] G.E. Mushi, G. Di Marzo Serugendo, P.-Y. Burgi. Digital technology and services for sustainable agriculture in Tanzania: A literature review. *Sustainability*, 14 (2022).

[5] X. Tian, B.A. Engel, H. Qian, E. Hua, S. Sun, Y. Wang. Will reaching the maximum achievable yield potential meet future global food demand? *Journal of Cleaner Production*, 294 (2021).

[6] M. Gebska, A. Grontkowska, W. Swiderek, B. Golebiewska. Farmer awareness and implementation of sustainable agriculture practices in different types of farms in Poland. *Sustainability*, 12 (2020).

[7] J. Chen, A. Yang. Intelligent agriculture and its key technologies based on Internet of Things architecture. *IEEE Access*, 7 (2019).

[8] M.S. Farooq, S. Riaz, A. Abid, K. Abid, M.A. Naeem. A survey on the role of IoT in agriculture for the implementation of smart farming. *IEEE Access*, 7 (2019).

[9] Ö. Köksal, B. Tekinerdogan. Architecture design approach for IoT-based farm management information systems. *Precision Agriculture*, 20 (2019).

[10] K.H. Law, K. Smartly, Y. Wang, "Sensor data management technologies". In: Wang, M.L., Lynch, J.P., Sohn, H. (eds.), "Sensor Technologies for Civil Infrastructures: Performance Assessment and Health Monitoring". Sawston, UK: Woodhead Publishing, Ltd. (in press).

[11] K. Smartly, K.H. Law, "Coupling wireless sensor networks and autonomous software for integrated soil moisture monitoring", The 10th International Conference Hydroinformatics. Hamburg, Germany, July 14, 2012.

[12] K. Smartly, K. Georgieva, M. König, K.H. Law, "Monitoring of slope movements coupling autonomous wireless sensor networks and web services", *The First International Conference on Performance-Based Life-Cycle Structural Engineering*. Hong Kong, China, December 5, 2012.

9 Crop Disease Detection Using Image Analysis

Mohamed Iqbal M., Ramireddy Jyothsna,
Somu Preethi Deekshita, and
Yaddanapudi Renuka Gayathri

9.1 INTRODUCTION

The basic supply of food, raw materials, and fuel that supports a country's economic growth is agriculture. As the world's population is expanding quickly, agriculture is finding it difficult to meet all its needs. Various challenges, including changes in the environment, diminishing inseminator populations, crop diseases, lack of proper irrigation, and more, persistently threaten global food security. The ability to control agricultural illnesses by spotting them when they show up on crops is a benefit. The cultivation of crops that are contaminated by illness is one of the biggest difficulties that farmers in rural areas encounter. Disease identification used to rely on agricultural extension organizations and local plant clinics, but now people can diagnose diseases online because of the heavy use of the latest technologies. Smartphones are uniquely equipped to aid in diagnosing diseases utilizing their computing power, active-matrix display, and built-in equipment. With a measured 5 – 6 billion smartphones world-wide by 2020 and 69% of the whole world having access to mobile internet connectivity at the end of 2015, digital connectivity has elevated 12-fold after 2007 [1]. Crop diseases have the potential to disrupt and harm both the ecological and economic equilibrium. Monitoring and identifying plant illness on farms is a time-consuming and labor-intensive task, demanding significant time and effort. [2].

Diagnosis of diseases in various plant parts holds great significance in today's world. To this end, image processing and machine learning (ML) offer a plethora of techniques that can be effectively harnessed for the analysis of images [3]. Plant diseases (like mosaic virus, late blight, tomato bacteria, early blight, and yellow curved) have recently become more prevalent and are negatively influencing plant progress along with quality and quantity of yield. Infections with viruses, bacteria, or fungi, for example, can be produced by insects or by any other element of nature. To achieve the right treatments and a better, quicker yield, these must be properly identified.

This conference paper aims to present a comprehensive exploration of crop disease detection using methodologies for analyzing images. This paper aims to introduce an application that, using leaf textural similarity, predicts the type of crop disease. The algorithm is trained using a dataset from a plant community that includes both healthy

 DOI: 10.1201/9781003484608-9

and damaged crop leaves. To stop additional crop damage, advanced identification of agricultural illness should be used, which is beneficial for maintaining farming.

In the following sections, we will review the key components of crop disease detection using image analysis. The paper will cover data collection and preprocessing techniques essential for obtaining high-quality images representative of different crop diseases and environmental conditions. Feature extraction methods will be discussed, as they play a crucial role in capturing unique disease-related patterns from the images.

9.2 LITERATURE SURVEY

Deep learning and machine learning techniques have led to significant advancements in crop disease detection. Convolutional Neural Networks (CNN) have exhibited significant promise in automatically extracting pertinent features from images and identifying diseases with high accuracy. The literature often discusses various image-based techniques for crop disease detection, including color-centric methods, texture examination, and form-based approaches. In the study of Naik et al. [4], chili viruses were classified using 12 different deep-learning models, including ResNet101, SqueezeNet, AlexNet, EfficientNetb0, DarkNet53, DenseNet201, InceptionV3, MobileNetV2, NasNetLarge, VGG19, ShuffleNet, and XceptionNet. The high-scoring algorithm was found to be VGG19 with a confidence value of 83.5%, excluding any data imputation techniques. In [5] it can be observed how the accuracy has increased when the architecture has changed from Alex Net 85.53% to Google Net 99.34%. After researching various approaches to recognize and identify plant diseases [6], it has been discovered that several popular architectures are currently being utilized. These include LeNet, AlexNet, VGGNet, GoogLeNet, InceptionV3, ResNet, and DenseNet, which have demonstrated a considerable increase in disease recognition accuracy. The implementation of these architectures has significantly contributed to the enhancement of crop disease recognition. In their study, Ghosal et al. [7] offered a deep CNN framework for the identification and categorization of eight unique types of soybeans. The authors introduced an interpretation method, which utilized top-K clear design maps to segregate the visual evidence and facilitate accurate calculations. For the identification of tea plant diseases from an image set having different leaves, an ideal CNN model called LeNet is developed [8], thereby enhancing the diagnostic evaluation of plant leaves. In their study, Dhaka et al. [9] presented an outline of the fundamental techniques employed in CNN models for disease detection through leaf images. The founders conducted a comparison of various Neural Network models, preprocessing methodologies, and frameworks. The findings of this research provide great info on the evolution of efficient, accurate plant disease identification systems using deep learning methods.

Hyperspectral and multispectral imaging are gaining popularity for crop disease detection due to their ability to capture a wide range of spectral information [10]. Images that have been enhanced have higher quality and clarity than the original image. The segmentation of the leaf picture is critical when extracting features from that image. In [11], in addition to image analysis, the model focuses on using other sensors, such as thermal cameras, to detect crop diseases based on temperature variations associated with infection. In [12], the use of hyperspectral and multispectral imaging, coupled with advancements in drone and Unmanned Aerial Vehicle (UAV) technology, has opened new avenues for disease detection. Prior to the manifestation of visible symptoms, these remote sensing techniques provide the opportunity for detecting diseases in their early stages, allowing for proactive control of illness and optimal resource allocation. Nagaraju et al. [13] conducted a thorough review to identify the most optimal datasets, data-cleaning approaches, and deep-learning methodologies for different plant species. The authors assessed 84 different research papers to evaluate the effectiveness of deep learning in disease diagnosis. According to new research, the recognition accuracy for a particular system has been reported to have reached 93.4%. In a similar study, Zhang et al. [14] proposed a technique to identify leaf diseases of cucumber which achieved a high accuracy rate of 94%.

9.3 PROPOSED METHODOLOGY

This part discusses the classification of the model into six classes using leaf images. A model was created to detect crop diseases in their initial phases, allowing for the implementation of targeted preventive measures before the diseases spread completely. The proposed model comprises five major components:

9.3.1 DATASET DESCRIPTION

The dataset utilized in this study is the open-source PlantVillage dataset, which was obtained by Kaggle [15]. The dataset comprises labeled images of diseased leaves, with 20,639 images of 256 x 256 pixels resolution, divided into 15 categories. In this study, only the initial six categories were considered for all experiments to achieve better accuracy. Figure 9.1 illustrates one image per category from the PlantVillage dataset. The batch size is set at 32, and the dataset is partitioned into training, testing, and validation sets. To prevent overfitting, the dataset is segmented into 80% for training, 10% for validation, and 10% for testing, since it is noteworthy that the model should perform well on unseen data. Table 9.1 provides additional information about the entire dataset.

9.3.2 DATA PREPROCESSING

Before constructing any model, it is imperative to preprocess the dataset. The TensorFlow package offers Keras preprocessing layers that enable the resizing and rescaling of dataset batches. Normalization is a technique that can facilitate the training of neural networks by rescaling the data such that it falls within the range of 0 and 1. The standard scaling can normalize almost every feature by reducing the mean (u) from the training data and fractionating it by the standard deviation (s).

Standard Scaling formula:

$$z = (x - u) / s \tag{1}$$

FIGURE 9.1 Visualization of images from each class.

TABLE 9.1
Brief Overview of the Dataset

Disease Class	Train	Test	Validation
Potato_Early blight	800	100	100
Potato healthy	122	15	15
Potato_Late blight	800	100	100
Tomato healthy	1273	159	159
Tomato_virus	299	37	37
Tomato Bacterial_spot	1702	212	212

This method (1) normalizes almost every feature by removing the average(u) of the trained data and dividing it by standard deviation(s). Data augmentation is also very useful for training neural networks. By applying this, more accurate predictions can be made on unseen data. Using Keras preprocessing layers for data augmentation can be done like layers. Random Flip and Layers. Random rotation over the same image repeatedly.

9.3.3 BUILDING A MODEL

To construct a model, a sequential CNN model with a convolution layer with dimensions (5,5), MaxPooling layer sized (2,2), nodes that have been flattened, and the dense layer's activation function being ReLU, which categorizes the six labeled objects, is used. Figure 9.2 shows an implementation of the CNN model.

9.3.4 MODEL ARCHITECTURE

Figure 9.3 shows a detailed explanation of the proposed model. This section provides a comprehensive and confident account of the model, as demonstrated in Figure 9.3. The input size is (32,256,256,3), and the dataset is efficiently divided into 32 batches. The first layer, being a Sequential Layer, creates no parameters, while the Conv2D layer generates an output of size (32, 256, 256, 32) and comprises 896 parameters. The following max_pooling layer does not create any parameters. After pooling, the flattening step is applied to streamline the model's connection. Two dense layers are utilized for flattening the layer.

The Suggested model's parameter count is approximately 278,086. Upon execution, the model delivers an impressive training dataset accuracy of 95.

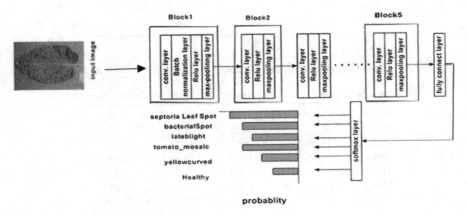

FIGURE 9.2   Implementation of CNN model.

```
model.summary()
```

Model: "sequential_6"

Layer (type)	Output Shape	Param #
sequential (Sequential)	(None, 256, 256, None)	0
sequential_1 (Sequential)	(None, 256, 256, None)	0
conv2d_20 (Conv2D)	(32, 254, 254, 32)	896
max_pooling2d_20 (MaxPooling2D)	(32, 127, 127, 32)	0
conv2d_21 (Conv2D)	(32, 125, 125, 64)	18496
max_pooling2d_21 (MaxPooling2D)	(32, 62, 62, 64)	0
conv2d_22 (Conv2D)	(32, 60, 60, 64)	36928
max_pooling2d_22 (MaxPooling2D)	(32, 30, 30, 64)	0
conv2d_23 (Conv2D)	(32, 28, 28, 64)	36928
max_pooling2d_23 (MaxPooling2D)	(32, 14, 14, 64)	0
conv2d_24 (Conv2D)	(32, 12, 12, 64)	36928
max_pooling2d_24 (MaxPooling2D)	(32, 6, 6, 64)	0
flatten_4 (Flatten)	(32, 2304)	0
dense_8 (Dense)	(32, 64)	147520
dense_9 (Dense)	(32, 6)	390

Total params: 278086 (1.06 MB)
Trainable params: 278086 (1.06 MB)
Non-trainable params: 0 (0.00 Byte)

FIGURE 9.3 Configuration of proposed CNN model.

9.3.5 MODEL TRAINING AND TESTING

The model has been subjected to rigorous training on both the training and validation datasets. A summary of the information on every epoch is presented in Table 9.2. It is noteworthy that the accuracy of the model has consistently improved with each epoch. Figure 9.4 illustrates the graphical comparison of Training vs Accuracy Validation and Training vs Loss Validation plots. The plot indicates that the training accuracy exhibited a regular increase over time, while the validation accuracy displayed some fluctuations.

TABLE 9.2
Accuracy and Loss for Each Epoch

Training loss for every epoch	Training accuracy for every epoch	Validation loss for every epoch	Validation accuracy for every epoch
0.9518	0.6374	0.5534	0.8109
0.3167	0.8848	0.3414	0.8832
0.2520	0.9127	0.6211	0.7845
0.2019	0.9239	0.3772	0.8651
0.1490	0.9461	0.3716	0.8832
0.1797	0.9339	0.7346	0.7796
0.1572	0.9443	0.1255	0.9572
0.1426	0.9485	0.2090	0.9276
0.1255	0.9541	0.4312	0.8849
0.1215	0.9571	0.4479	0.8766
0.0938	0.9643	0.2068	0.9309
0.0821	0.9653	0.3154	0.8997
0.0770	0.9748	0.1762	0.9359
0.0954	0.9681	0.2742	0.9161
0.0837	0.9685	0.3876	0.8832
0.0814	0.9710	0.1503	0.9474
0.0899	0.9671	0.1042	0.9605
0.0941	0.9675	0.6306	0.8224
0.0856	0.9688	0.2319	0.9276
0.0603	0.9772	0.1458	0.9507

9.4   RESULTS

After the experimentation on training the model, it is tested and evaluated. On evaluation, it is observed that the accuracy is 96.5 on the testing dataset which is the best accuracy. Figure 9.5 shows the definite label, the model-estimated label, and the confidence value of the images.

Figure 9.4 depicts the contrast in Loss and Accuracy between the Trained and Validation sets. It is observed that using TensorFlow and Keras, image classification is done with an accuracy of 95%, which shows that the model performed best on unseen data.

9.5   CONCLUSION

In conclusion, crop disease detection using image analysis has emerged as a promising and effective approach to safeguarding agricultural productivity and ensuring food security. The combination of deep learning, machine learning, and advanced analysis of image techniques has changed the way we recognize and combat plant diseases, offering significant advantages over traditional manual methods. This technology holds the promise of revolutionizing agricultural practices through its

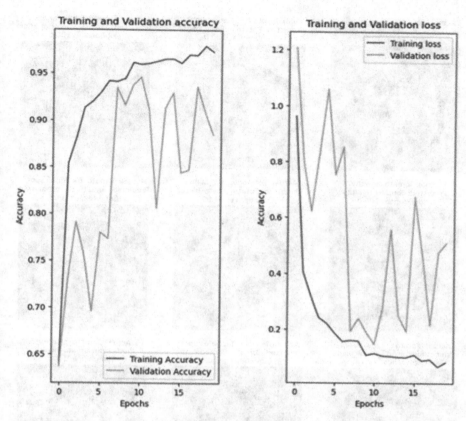

FIGURE 9.4 Plots depicting accuracy and loss.

capability to provide early and precise disease diagnosis. As a result, it can minimize yield losses and diminish dependence on agrochemicals.

Recent advancements in crop disease management have been primarily facilitated by the application of ML algorithms, especially deep learning techniques, such as CNN. These neural networks demonstrated remarkable potential by efficiently learning significant features from images, leading to highly accurate disease identification. In conclusion, the integration of image analysis techniques with advanced technologies can greatly transform crop disease management. By stimulation and precise disease detection, this approach can significantly contribute to sustainable agriculture, mitigate economic losses, and enhance global food security. While exploration continues to progress in this field, it is crucial to disseminate knowledge, promote technology adoption, and ensure that the advantages of crop disease detection using image analysis are accessible to farmers worldwide.

FIGURE 9.5 Actual and predicted class of different images and confidence value.

REFERENCES

1. ITU (2015). ICT Facts and Figures – The World in 2015. Geneva: International Telecommunication Union.
2. T. A. Salih, A. J. Ali, and M. N. Ahmed (2020). Deep learning convolution neural network to detect and classify tomato plant leaf diseases. *Open Access Library Journal*, 7, 1–12, doi: 10.4256/oalib.1106296
3. M. E. Pothen and M. L. Pai (2020). "Detection of Rice Leaf Diseases Using Image Processing," 2020 Fourth International Conference on Computing Methodologies and Communication (ICCMC), Erode, India, pp. 424–430, doi: 10.1109/ICCMC48092.2020.ICCMC-00080

4. B. N. Naik, M. Ramanathan, and P. Palanisamy (2022). Detection and classification of chili leaf disease using a squeeze-and-excitation-based CNN model. *Ecological Informatics*, 69, 101663.

5. S. P. Mohanty, D. P. Hughes, and M. Salathé (2016). Using deep learning for image-based plant disease detection. *Frontiers in Plant Science*, 7, 1419.

6. A. Abade, P. A. Ferreira, and F. de Barros Vidal (2021). Plant diseases recognition on images using convolutional neural networks: A systematic review. *Computers and Electronics in Agriculture*, 185, 106125, ISSN 0168-1699.

7. S. Ghosal, D. Blystone, A. K. Singh, B. Ganapathysubramanian, A. Singh, and S. Sarkar (2018). An explainable deep machine vision framework for plant stress phenotyping. *Proceedings of the National Academy of Sciences USA*, 115(8), 4613–4618.

8. S. Gayathri, D. C. J. W. Wise, P. B. Shamini, and N. Muthukumaran (2020). "Image Analysis and Detection of Tea Leaf Disease Using Deep Learning," 2020 International Conference on Electronics and Sustainable Communication Systems (ICESC), Coimbatore, India, pp. 398–403, doi: 10.1109/ICESC48915.2020.9155850

9. V. S. Dhaka, S. V. Meena, G. Rani, D. Sinwar, M. F. Ijaz, and M. Woźniak (2021). A survey of deep convolutional neural networks applied for prediction of plant leaf diseases. *Sensors*, 21, 4749.

10. S. D. Khirade and A. B. Patil (2015). "Plant Disease Detection Using Image Processing," 2015 International Conference on Computing Communication Control and Automation, Pune, India, pp. 768–771, doi: 10.1109/ICCUBEA.2015.153

11. J. Zhang , Y. Huang, R. Pu, P. Gonzalez-Moreno, L. Yuan, K. Wu, and W. Huang (2019). Monitoring plant diseases and pests through remote sensing technology: A review. *Computers and Electronics in Agriculture*, 165, 104943.

12. N. Kitpo and M. Inoue (2018). "Early Rice Disease Detection and Position Mapping System Using Drone and IoT Architecture," 2018 12th South East Asian Technical University Consortium (SEATUC), Yogyakarta, Indonesia, pp. 1–5, doi: 10.1109/SEATUC.2018.8788863

13. M. Nagaraju and P. Chawla (2020). Systematic review of deep learning techniques in plant disease detection. *International Journal of Systems Assurance Engineering and Management*, 11, 547–560.

14. L. Li, S. Zhang, and B. Wang (2021). Plant disease detection and classification by deep learning – A review. *IEEE Access*, 9, 56683–56698, doi: 10.1109/ACCESS.2021.3069646

15. PlantVillage dataset. Available on www.kaggle.com/datasets/emmarex/plantdisease

10 Automated Detection of Plant Diseases Utilizing Convolutional Neural Networks

*Archisman Panda, Arjav Anil Patel,
Muhammad Zakaria Shaik, Sandeep Kumar
Satapathy, Sung-Bae Cho, Shruti Mishra,
Abishi Chowdhury, and Sachi Nandan Mohanty*

10.1 INTRODUCTION

With agriculture making up more than 25% of Gross Domestic Product (GDP), it is an important economic sector for emerging nations. Pests and crop diseases, however, have a major effect on food security and agricultural output. Precision agriculture (PA) techniques [1] have been used to boost yields and efficiency in response to these issues. Conventional techniques, including visible inspections, are more expensive, labor-intensive, and less precise. In recent decades, non-invasive approaches like image processing have become more popular. Crop data is captured by cameras equipped with sensitive sensors, and machine learning algorithms are then utilized to process the resulting photos. Small datasets and manually constructed feature extraction techniques are the main drawbacks of traditional machine-learning approaches. With developments in graphics processing units (GPU) and computer vision, deep learning is seen to be a useful tool for improving automated processes.

Plant diseases are the only thing keeping 70% of India's farmers from losing crop yields. Manually observing plant diseases [2] is difficult since it requires time, labor, and skill. Using photographs of leaves, image processing and machine learning models can be used to detect diseases [3][4][5]. While machine learning is a branch of artificial intelligence that comprehends training data and incorporates it into models for decision-making, image processing is the process of extracting meaningful information from images. Color, damage, area, and texture criteria are used in classification. The study aims to determine plant leaf diseases as accurately as possible by analyzing several image data. Plant disease identification was previously accomplished by experts using chemical procedures or visual inspection, both of which were costly and time-consuming.

DOI: 10.1201/9781003484608-10

Promising outcomes have been observed in the accurate identification of plant diseases from digital photos through the application of machine learning (ML) and deep learning (DL) techniques. Conventional ML [6] methods [7], such as featuring extraction and classification, are commonly employed to identify disease symptoms resulting from abiotic stressors like drought and nutrient shortage, as well as diseases like rust, powdery mildew, and leaf blotch. However, the ability to precisely recognize subtle symptoms and detect diseases in their early stages is limited by these methods. Research studies have proposed different ML and DL approaches for plant disease detection [8][9], but most focus on specific types of diseases or plant species. More research is needed to develop a generalizable and robust model that can work for different plant species and diseases. Transfer learning and ensemble methods have emerged as popular trends in plant disease detection using ML and DL.

Extending their research to include 14 crops and 26 diseases in 2016, Mohanty et al. achieved a 99.35% accuracy rate on [10] the reserved test set. While Zhang et al. enhanced deep convolutional generative adversarial networks (DCGAN) for the purpose of detecting faults in pear photos, Xu et al. presented a method for augmenting data. Five CNN models were trained by Yasamin Borhani et al., with EfficientNet-B0 exhibiting the highest accuracy rate in scenarios involving inexpensive computation. To combat plant disease detection, techniques such as Transformer and YOLO have been employed in addition to CNNs. Although Liu and Wang made improvements to the YOLOv3 model-based tomato disease image detection technique, YOLOv3 still has certain limitations. By using it on plants, Midhun P. Mathew et al. were able to establish a simple and effective plant disease detection method.

To feed the growing global population, agricultural food production must rise by at least 70% on a worldwide scale. But the weather, the soil, and irrigation water—all of which are outside of human control—are crucial to the agricultural industry. Drones and other [11] precision technologies are essential for increasing agricultural productivity and optimizing resource utilization. Pest identification, crop yield forecasting, crop spraying, water stress detection, land mapping, nutrient deficiency diagnosis, weed detection, livestock control, agricultural product protection, and soil analysis are just a few of the uses for drones. Drones are a valuable tool in the fight against crop loss since they can detect and prevent plant diseases early on. Drone-based decision-support systems have the potential to save labor costs, boost production, enhance product quality, and facilitate better decision-making.

10.2   RELATED WORK

Esmaeil Najafi et al. [12]. in their paper, discussed how plant diseases impact productivity and proposed a DL approach using the Vision Transformer (ViT) architecture. The study presented a DL approach that primarily utilizes the Vision Transformer (ViT) architecture to classify plant diseases in time automatically. Different model variations, such as ViT-based methods, Convolutional Neural Network (CNN) approaches, and a combination of CNN ViT models, were examined by the author. The researchers evaluated various model modifications to provide farmers with the right information for preventive actions, emphasizing the trade-off between accuracy and speed. The ultimate objective is to provide farmers with information for preventive measures. In their paper Jackulin C. et al. [13]. Have provided a comprehensive overview of several ML and DL techniques for plant disease diagnosis. It emphasizes a data-driven approach and recognizes the effectiveness of DL models compared to traditional ML. The author has emphasized the important role of agriculture in India and the need to increase crop yields to meet the growing food demand. The focus is on solving plant disease detection approaches, especially ML and DL, to combat low yields caused by diseases. The importance of data-driven ML in disease detection was kept in mind, and a comprehensive overview of different methods for artificial intelligence-based plant disease detection was provided. The result highlights the importance of extensive research on machines and DL methods for plant disease recognition to help farmers through automatic disease detection. The study recognizes advances in scaling models and identifies research gaps that need to be addressed for more effective plant disease detection.

Research shows that DL models are more effective than ML. A method that uses scanning to detect plant diseases in a computational scenario was introduced by Chunli Lv in her paper [10]. The method involved collecting data sets that adjust network parameters to improve accuracy and use dynamic inference to adapt the network structure for different device platforms. This study demonstrates efficiency on a high-performance computing platform and demonstrates practical utility in real-world scenarios. Tests show that the model is efficient on high computing platforms, reaching a speed of 58 FPS and a high accuracy of 0.94. This approach can be used effectively in real-world situations where it demonstrates practical utility. The advancements achieved in detecting plant diseases using ML and DL technologies were emphasized by Muhammad Shoaib by DL techniques like CNNs [6]. While acknowledging these improvements, the author also points out some limitations, such as the reliance on studies conducted in laboratories and the importance of having datasets. The collaboration between agriculture experts and DL specialists is considered crucial. Challenges related to data availability and computational costs are recognized. The paper encourages researchers to develop models that can be applied broadly and publicly accessible datasets, providing insights for practitioners interested in advancements and challenges associated with plant disease detection. This paper by Mohanty et al. [14] explores the potential of using smartphones equipped with learning-based computer vision for crop disease diagnosis. By utilizing a dataset composed of 54,306 images showcasing both healthy plant leaves, the author trains a CNN that can accurately identify 14 different crop species and 26 diseases.

This impressive model achieves an accuracy rate of 99.35% when tested. The study emphasizes the opportunities for deploying smartphone-assisted crop disease diagnosis due to the use of smartphones worldwide. While this approach demonstrates the feasibility, it acknowledges the importance of having diverse training data and expanding perspectives in the image collection. It aims to supplement existing methods rather than replace them altogether. The paper envisions the future role of smartphones in crop loss prevention, including existing methods. The research by Radhika Bhagwat et al. extensively investigates methods used for detecting plant diseases with a focus on imaging techniques and processing algorithms [1]. It compares ML approaches with learning techniques when analyzing images in both visible and spectral ranges. The evaluation considers aspects such as datasets, processing methods, models, and efficiency. The results indicate that DL when using CNN models, outperforms ML in terms of achieving high accuracy across different plant species and diseases. However, there are challenges involved, including the requirement for datasets and limitations in the availability of accessible data. In addition to discussing the popularity of Red Green Blue (RGB) imaging, the study also highlights the limitations of detecting diseases before symptoms appear. Suggests the benefits of using hyperspectral and multispectral imaging techniques.

Despite these challenges, the research envisions a future where smartphones equipped with sensors can be utilized along with compact DL algorithms to enable time and accurate early detection of diseases in plants. The use of drones in detecting plant diseases as a way to address the challenges faced by the industry was suggested by Ruben Chin et al. [11]. The focus is on how diseases affect both the quality and quantity of crops. Through a review of existing literature, 38 primary studies are identified. The common disease found is blight, with fungi being the culprit. Quadcopters are the type of drone used for this purpose utilizing color infrared images and CNNs for classification tasks. Grape and watermelon crops have been extensively studied, while kiwi, squash, pear, lemon, onion, and rice offer avenues for research. The article emphasizes the importance of conducting investigations, particularly for lesser-known disease categories, and highlights the promising role that drone technology can play in agriculture. The growing importance of deep learning (DL) in addressing plant diseases, which pose a threat to food security, was explored by Rokhman et al. [15]. They reviewed 40 studies that applied DL techniques to agriculture and found that DL outperforms image processing methods. The authors suggested areas for future research, including advanced techniques such as drones, agricultural robots, and Generative Adversarial Networks (GAN).

The paper by Khaleel Khan et al. [16]. discussed the shift from ML to DL in plant disease detection and highlighted the importance of early disease detection through advanced technology. The study focused on using technology, specifically artificial intelligence (AI), with a focus on ML and DL, to automate the identification of diseases in plants. The authors have discussed how there has been a shift from ML approaches to learning methods over the past five years. They also delve into datasets related to plant diseases. The paper highlights the challenges faced by systems, Emphasizing the importance of timely disease detection through advanced technologies. In conclusion, the paper emphasizes how ML and DL play a role in identifying plant diseases as well as their impact on crop yield, global warming, and potential famine if not effectively addressed. This research showcases how computer vision techniques are gaining acceptance in this field and suggests avenues for exploration and improvement in automating plant disease identification. Kulkarni et al. [2] outline a system that uses computer vision to detect plant diseases in India, a country where agriculture plays a role for 70% of the population. Traditional methods of identifying plant diseases involve labor. Consume a significant amount of time.

However, the proposed solution employs image processing and ML models to analyze images of leaves, taking into account factors like color, damage, area, and texture.

The system achieves an accuracy rate of 93% and an F1 score of 0.93, demonstrating its effectiveness when compared to other approaches. Noteworthy advantages include computing capabilities, affordability, and the ability to monitor crop fields without requiring specialized equipment. In the paper by S. Rajagopal et al. [17], the authors propose an automatic plant leaf disease detection and classification system using a cellular network-based CNN (OMNCNN). This model combines different steps, including preprocessing, segmentation, using MobileNet, and classification using an extreme learning machine (ELM). Hyperparameters were optimized using Imperial Penguin Optimizer (IPO). The proposed OMNCNN model outperforms state-of-the-art methods in plant leaf disease detection and classification with high accuracy, recall, precision, F-score, and kappa value. The paper highlights the importance of automated methods in overcoming the challenge of hand disease detection in agriculture, especially in India, where 35% of crop yields are lost due to plant diseases. The paper by B.Vijayalakshmi et al. [18] presented a Parallel and Distributed Simulation Framework (PDSF) integrated with the Internet of Things (IoT) for smart agriculture with pest control and agricultural monitoring. The proposed system reduced the workload on a single-core GPU by distributing tasks among multiple GPUs and ensuring data delivery even in case of failure. The four levels of the system (crop management, pest control, production activities, and input functional areas) were efficiently processed simultaneously, resulting in a significant increase of 98.65%. The research focused on the feasibility of combining complex simulation with information and communication technology (ICT) for advanced agricultural production, decision-making and data analysis, nutrient analysis, and plant condition control experiments.

The paper by Ngongoma et al. [19]. focuses on the differences between developed and developing regions in terms of agriculture. This study highlights the need to introduce agricultural practices, especially in Africa, to address food insecurity, which is exacerbated by population growth. Plant diseases show areas that still need improvement, such as research, real-time monitoring, and effective actions to reduce impacts. The paper concludes by providing additional recommendations to improve the accuracy and efficiency of the model, which forms the basis for research, especially in future doctoral research projects. The growing field of agricultural robots was explored by Ampatzidis et al. [20]. with a particular focus on applications and challenges in robotic plant safety and disease detection. An introduction was provided to the growing interest and effort in agricultural robotics research, especially in automating routine and time-consuming tasks. The article explores the importance of accurate crop protection and environmental control in agricultural sustainability, Discussing the potential of robotic applications in disease detection and emphasizing the need for diagnostic features when challenges arise, such as distinguishing between diseased and healthy plants. The author also touches on the environmental and social sustainability aspects of robotic plant protection, considering factors such as chemical emissions and operational impact. In conclusion, the authors acknowledged the progress made by highlighting ongoing challenges in disease diagnosis and called for further exploration of the role of image analysis in robotic management. The potential of robots as skilled pathologists and the importance of incorporating rapid molecular diagnostic methods into vision-based robots for diagnostic specificity are discussed.

Overall, the authors call for an integrated approach to address the growing field of agricultural robots and improve the capabilities of robot observers and analysts. A review of conventional processes used in plant disease detection, highlighting the shortcomings of subjective, time-consuming methods that rely on plots and artificial feature extraction, was conducted by Li et al. in their paper [21].

This paper reviewed the progress made by DL in plant disease recognition, emphasizing transparency, automatic extraction, and the key role of larger databases, GPU adaptation, and specialized software libraries. CNNs are emerging as a method of interest, especially when combined with transition learning to adapt pre-trained models for plant recognition. The gaps identified in the review of the existing literature, especially recent changes in visualization techniques and the lack of adaptation to DL models, lead to the aim of the paper. It aimed to fill this gap through comprehensive research in plant leaf disease recognition, image processing, hyperspectral imaging, and DL. The paper covered the background knowledge of DL, existing studies on the application of DL in crop foliar disease detection, challenges and solutions for plant disease detection with limited databases, and hyper-research. It is a summary of spectral imaging applications and highlights gaps in current literature that require attention. In conclusion, given sufficient data, it is confirmed that the DL method shows high accuracy in the recognition of plant leaf diseases. This highlights the importance of developing reliable models adapted to different disease databases and calls for the creation of large databases in real-world settings. The difficulty in obtaining a specific database for detecting early plant diseases using hyperspectral imaging is known, indicating that this problem must be addressed for widespread adoption. The paper by Andrea Luvisi et al. [22] outlined the recent advances in agricultural robotics research with potential applications and manufacturing efforts in agricultural robotics. The focus was on automating routine farming tasks, including operations such as land preparation, planting, fertilizing, and harvesting. Robots excel in continuous, repetitive, and time-dependent tasks, especially in agricultural and temporary crop situations.

However, the complexity of robotic plant protection increases because it involves not only physical challenges but also important aspects of pathogen diagnosis. It highlights the importance of technological capabilities such as vision, machine vision, global positioning systems, laser technology, actuators, and mechatronics in the development of agricultural robots. It also recognizes the impact of the growing online landscape, including the IoT and cloud-based solutions, creating unique opportunities for automated and robotic systems in agriculture. In particular, the integration of IoT and Wireless Sensor Networks (WSN) into plant control systems is highlighted. ML, machine vision, and AI are taking center stage in agricultural applications, contributing to the automation of routine tasks and expanding to detect and diagnose plant diseases. Despite these advances, the review explores the challenges associated with the detection of plant diseases. Successful disease detection methods often require a controlled environment for data acquisition to reduce false positives, which is a limitation of current approaches. The key role of image analysis and recognition in automatic crop protection is discussed in detail. This paper raises questions about the suitability of image analysis as a key tool for plant monitoring, diagnosis, and management, especially in the case of vegetable and fruit crops. This highlights the need for close communication between the development of robotic systems and diagnostic systems, as advanced image analysis methods for plant pathogen recognition are sometimes developed independently of robotic applications. In conclusion, the review suggests future directions for research and development in agricultural robotics. Improve fleet management and coordination among robots specialized in different tasks. In addition, in-situ detection tests such as Loop-Mediated Isothermal Amplification (LAMP) and Electrical Impedance Spectroscopy (EIS)-based lab-on-a-chip are proposed for rapid molecular diagnosis in the field. In conclusion, It provides insights into the current state of agricultural robots by addressing technological trends, challenges, and possible avenues for future research in the dynamic field of crop protection and disease management.

10.3 METHODOLOGY

10.3.1 DATA COLLECTION

Collecting a dataset for automated plant disease monitoring is a key step to ensure the efficiency and reliability of the proposed system [23]. From the careful selection of species and plant diseases, a deep understanding of individual diseases is essential for diagnosis and control. Datasets are important for collecting unique disease symptoms in different crops. One of our project's aims is to obtain the broad and diverse datasets needed to improve model generalization. It is essential that the dataset accurately reflects real-world scenarios, including plant growth stages, light conditions, and disease severity. Strict quality control measures, such as removing low-quality or irrelevant images and ensuring tag consistency, demonstrate the commitment to maintaining dataset integrity. Using data augmentation techniques such as rotation, shift, and zooming, we have augmented the dataset and increased the power of the model.

10.3.2 DATA PREPROCESSING

This process checks for the quality of images and changes in them; even the rotation and size of images impacts the dataset. In this case, the leaves were viewed from each angle to check whether they had a disease or not. Preparing data is essential for computer vision systems. Background noise needs to be eliminated in order to obtain accurate results. Greyscale images are obtained by converting RGB images, smoothing them with a Gaussian filter, and producing binaries by applying Otsu's thresholding algorithm. To enhance data quality and relevance, the AI model was preprocessed using techniques like image resizing and pixel value normalization. In order to address class imbalance and guarantee a representative distribution of diseases, oversampling techniques were applied. By adding random rotations, flips, and zooms to the data, the training set's diversity was enhanced, which enhanced the model's robustness and generalization in the identification of diseases. Shape, texture, and color features were extracted from the segmented leaf image using the bitwise AND operation.

10.3.3 TRAINING THE AI MODEL FOR INDIVIDUAL CLASS OF LEAVES

Our model training methodology involved separating the data into training and validation sets. We used a two-step process, first pre-training the chosen neural network architecture on the ImageNet dataset, then fine-tuning it on our plant disease images. We tuned hyperparameters like learning rates, dropout, and batch size to find the right balance between model complexity and ability to generalize. To evaluate model performance, we utilized a held-out test set not used during training. We measured key evaluation metrics such as accuracy, precision, recall, and F1 score. We also did cross-validation across different data splits to test consistency. In short, our approach combines a carefully labeled dataset, robust preprocessing, a well-justified neural network, and systematic training, validation and testing. This methodology ensures a robust and reliable automated plant disease detection system.

Leaf detection is done first by training our model on a classifier dataset containing elements of each leaf. The model developed for leaf identification makes a significant contribution to the next stage of disease diagnosis. Creating the optimal model is an important aspect of our automatic disease control approach. Using CNNs for their proven power in automatic learning of hierarchical features, we carefully prepared the data by dividing the labeled dataset into training and validation sets. Ensuring a balanced distribution of healthy and diseased samples is an integral part of our approach. For conventional ML models, feature extraction involves extracting relevant features such as color histograms and texture features from an image. The knowledge of the configuration of the model architecture, which defines the layers, activation functions, and input/output parameters, is very important for DL models such as CNNs. Model development is an iterative process that requires adjusting parameters to optimize performance. Each and every class is trained individually to identify its species.

10.3.4 ALGORITHMIC PROCESS FOR TRAINING THE AI MODEL FOR DISEASE DETECTION

The model first imports the packages for resizing, rescaling, and data augmentation from tensorflow.keras.preprocessing library. Data is prepared by dividing the dataset into training, validation, and testing datasets and then resized, rescaled, and augmented with the help of the tensorflow libraries. The model is created using basic CNN layers as the base model, and then adding appropriated dropout, Flatten and Dense layers to maximize accuracy. The model summary shows a total of 21 million parameters. The model is then compiled using the Adam optimizer and SparseCategoricalCrossentropy losses. Then, the 13 models are successively trained on the number of epochs that give maximum accuracy, observed through trial and error. Each model is trained for maximum accuracy and minimum loss, and then the test dataset is evaluated to validate the training accuracies and ensure overfitting has not occurred.

10.3.5 OpenMPI Implementation

Finally, the models thus created are integrated together. The steps of leaf detection and disease detection are integrated through the OpenMPI operations used on the trained models. OpenMPI divides the 13 models into three processors and integrates them simultaneously, resulting in a much better performance than the original performance using only one processor. This parallelization function is an integral concept proposed by this research. Thus, both DL and parallel processing are equally significant in the proposed model.

10.4 PROPOSED MODEL

As given in Figure 10.1, the proposed model utilizes the 13 datasets we made for each plant in the original dataset. Each dataset is resized and rescaled, then extensively augmented through the use of tensorflow.keras.layers.experimental.preprocessing classes. The backbone convolutional network consists of multiple sets of convolution-pooling layers, each consisting of two convolution layers with increasing kernel size followed by a max pooling layer. The convolution layers are followed by a Flatten layer and two Dense layers to compress the tensor into the number of classes in each respective dataset. The model is trained using appropriate epochs and evaluated using a testing dataset. After each plant has been trained, a small dataset containing images of each plant is trained on the code and evaluated. Lastly, all the models are integrated using OpenMPI operations.

10.5 RESULT ANALYSIS

Table 10.1 showcases the training, validation, and testing accuracy of each plant, as well as the classifier model that determines which plant the leaf belongs to. The proposed model has achieved high accuracy even in testing and has accurately (Figure 10.2) predicted leaf diseases of a test prediction batch of 100 leaves. A custom CNN model is most efficient for this dataset because pre-trained models give relatively lower accuracy due to the sheer size as well as the nature of augmentation of the dataset. Each dataset has been pre-augmented as well as extensively augmented before training to ensure maximum accuracy. This model can be utilized extensively and generally for all plants, hence the pressing need for OpenMPI.

OpenMPI is essential for the working of this model due to the massive size of the dataset as well as the extent of the purpose of the model. This model can be used to detect the diseases of any plant as long as we have a dataset of training images. Hence, to further generalize the model, an even more versatile dataset will be required. Hence, openMPI operations are required to distribute the load of training such a huge dataset onto multiple processors, which will significantly reduce the time required for training. All the processors in the device will simultaneously train our model for the

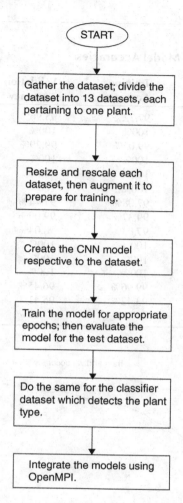

FIGURE 10.1   The flowchart representing the proposed work.

huge dataset, thus decreasing the time taken. This project emphasizes the need for openMPI in ML.

Some of the datasets have achieved 100% accuracy due to limited classes of data for that particular plant. In our future work, we plan to add more classes to detect the diseases of the plants (Figure 10.3). We have checked the accuracy and losses of each plant and plotted them in graphs. For the apple plant, training with ten epochs resulted in the model being overfitted and the test accuracy being unacceptable. Hence, nine epochs was the optimal solution. In plants like cherry and pepper bell, training on epochs above ten gives nearly similar accuracies due to the accuracy being in close proximity to 100%. The tomato plant dataset has a significantly lower accuracy than the plants with fewer classes; however, it still has a high accuracy of 93% after 15

TABLE 10.1
Comparative Analysis of Model Accuracies

Model	Epochs	Training Accuracy	Validation Accuracy	Testing Accuracy
Apple	9	97.32%	95.96%	96.35%
Blueberry	2	100%	100%	100%
Cherry	10	99.03%	99.29%	98.86%
Corn	2	100%	100%	100%
Grape	2	100%	100%	100%
Orange	2	100%	100%	100%
Peach	10	97.08%	97.44%	98.01%
Pepper Bell	11	99.37%	99.09%	99.22%
Potato	10	97.03%	98.04%	97.98%
Raspberry	2	100%	100%	100%
Soyabean	2	100%	100%	100%
Squash	2	100%	100%	100%
Strawberry	11	99.96%	96.45%	97.44%
Tomato	15	93.42%	95.41%	94.56%
Classifier	25	91.14%	93.12%	91.26%

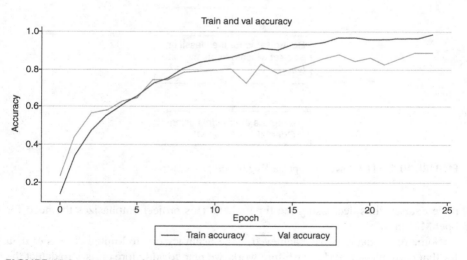

FIGURE 10.2   Accuracy of classifier dataset.

epochs. The classifier dataset had been trained with up to 50 epochs; however, 25 epochs have been found to give the highest accuracy due to being an optimal link between fewer epochs with lower accuracy and more epochs with higher accuracy. Each of the epochs trained and tested has been found to be the peak for that plant after plotting the training and validation losses on a graph.

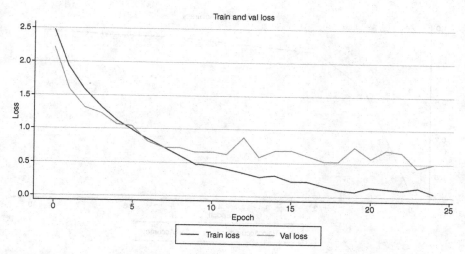

FIGURE 10.3 Loss of classifier dataset.

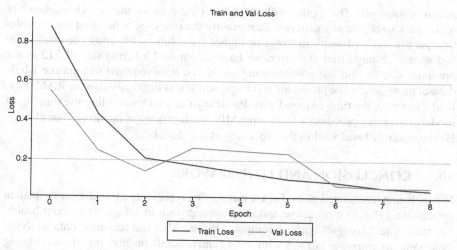

FIGURE 10.4 Loss graph of potato.

Through trial and error methods, we have found the model that gives the highest accuracy for the plant disease dataset. Multiple convolutional layers have been changed, and Dense layers and dropout layers added to prevent overfitting of the model. The models have been trained with multiple number of epochs and the epochs giving maximum accuracy have been chosen for the models. The accuracy (Figures 10.4 and 10.5) will decrease if layers are removed or if the Dense layer size is increased or decreased.

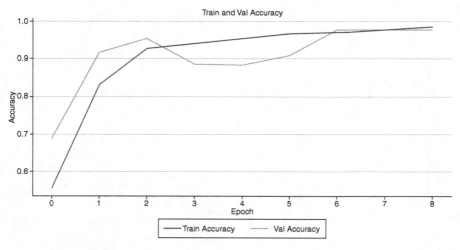

FIGURE 10.5   Accuracy graph of potato.

Ablation Analysis: The model will not work if we remove the last Dense layer or make the kernel size any different than exactly the one used. Changing the convolution parameters will result in lower or higher convolution, affecting the model size and speed. Through trial and error, we have determined a kernel size of 512 as the optimum size for the last convolutional layer. The time required to train the whole dataset using solely one processor was huge, and it was impossible due to RAM overload. However, the time required was drastically reduced by dividing the training of models into three processors using openMPI, each processor handling four classes. Hence, openMPI was vital in the efficiency of the model.

10.6   CONCLUSION AND FUTURE WORK

In conclusion, automated plant disease monitoring represents a transformative leap in agriculture, offering a proactive and efficient approach to safeguarding crop health. Leveraging technologies such as computer vision, ML, and real-time data analysis, these systems empower farmers with timely and accurate insights into the well-being of their crops. The ability to swiftly detect and diagnose diseases enables proactive interventions, preventing widespread crop damage and reducing reliance on traditional, often reactive, methods. These automated monitoring systems contribute significantly to sustainable agriculture by optimizing resource use and minimizing the need for indiscriminate pesticide applications. The seamless integration of advanced algorithms with imaging devices provides a scalable solution adaptable to diverse crops and geographical regions. Moreover, the continuous evolution of these systems, marked by regular model updates, data enhancements, and performance monitoring, ensures their resilience in the face of changing environmental conditions and emerging plant diseases. As technology continues to advance, automated plant disease monitoring not only enhances agricultural productivity but also plays a pivotal role in global food security. By fostering a harmonious synergy between human expertise and cutting-edge technology, these systems underscore the potential to revolutionize how we approach crop management, ushering in a new era of PA that is both sustainable and responsive to the dynamic challenges faced by farmers worldwide.

In essence, automated plant disease monitoring is like a watchful guardian for our crops, bringing a touch of technology to the fields. Imagine having a vigilant ally that sees what we might miss – the subtle signs of plant distress that hint at potential diseases. It is not just about spotting a problem; it is about doing so before it turns into a crisis. These systems act like a farmer's personal detective, using smart cameras and intelligent algorithms to observe the fields around the clock. They are not only looking out for the red flags of diseases but also whispering insights into the farmer's ear, allowing for timely and targeted interventions. It is the difference between a carefully crafted strategy and a reactive scramble when issues arise. Picture a scenario where farmers can focus on the art of farming – tending to their crops with care – while technology takes care of the early warning signs. Automated monitoring is not just about efficiency; it is about empowering farmers to make informed decisions that can save their harvests. Moreover, these systems are not static; they are dynamic partners in the journey of agriculture. They grow and adapt, learning from each season and getting smarter with every update. This human-touch technology embodies a sustainable future where the harmony between nature and innovation fosters healthier crops, more resilient farms, and a brighter outlook for global food security. Automated plant disease monitoring is more than a tool; it is a bridge between tradition and progress, ensuring that the age-old craft of farming thrives in our modern world.

In our future work, we aim to generalize the model for even bigger datasets with more plants, and also create an automated monitoring system to monitor the crops. We plan to propose and implement an automated alert system to alert the user in case of any anomalies observed by the model. We also plan to simulate a model data drift mechanism to ensure that the model keeps giving high accuracy.

REFERENCES

[1] Bhagwat, R., & Dandawate, Y. (2021). A review on advances in automated plant disease detection. *International Journal of Engineering and Technology Innovation*, 11(4), 251.

[2] Kulkarni, P., Karwande, A., Kolhe, T., Kamble, S., Joshi, A., & Wyawahare, M. (2021). Plant disease detection using image processing and machine learning. arXiv preprint. arXiv:2106.10698

[3] Iqbal, Z., Khan, M. A., Sharif, M., Shah, J. H., ur Rehman, M. H., & Javed, K. (2018). An automated detection and classification of citrus plant diseases using image processing techniques: A review. *Computers and Electronics in Agriculture*, 153, 12–32.

[4] Cap, Q. H., Uga, H., Kagiwada, S., & Iyatomi, H. (2020). Leafgan: An effective data augmentation method for practical plant disease diagnosis. *IEEE Transactions on Automation Science and Engineering*, 19(2), 1258–1267.

[5] Fiona, J. R., & Anitha, J. (2019, February). Automated detection of plant diseases and crop analysis in agriculture using image processing techniques: A survey. In 2019 IEEE International Conference on Electrical, Computer and Communication Technologies (ICECCT) (pp. 1–5). IEEE.

[6] Shoaib, M., Shah, B., Ei-Sappagh, S., Ali, A., Ullah, A., Alenezi, F., ... & Ali, F. (2023). An advanced deep learning models-based plant disease detection: A review of recent research. *Frontiers in Plant Science*, 14, 1158933.

[7] Wang, G., Sun, Y., & Wang, J. (2017). Automatic image-based plant disease severity estimation using deep learning. *Computational Intelligence and Neuroscience*, 2017(2917536), 8. https://doi.org/10.1155/2017/2917536

[8] Saleem, M. H., Potgieter, J., & Arif, K. M. (2021). Automation in agriculture by machine and deep learning techniques: A review of recent developments. *Precision Agriculture*, 22, 2053–2091.

[9] Waghmare, H., Kokare, R., & Dandawate, Y. (2016, February). Detection and classification of diseases of grape plant using opposite colour local binary pattern feature and machine learning for automated decision support system. In 2016 3rd International Conference on Signal Processing and Integrated Networks (SPIN) (pp. 513–518). IEEE.

[10] Liu, Y., Liu, J., Cheng, W., Chen, Z., Zhou, J., Cheng, H., & Lv, C. (2023). A high-precision plant disease detection method based on a dynamic pruning gate friendly to low-computing platforms. *Plants*, 12(11), 2073.

[11] Chin, R., Catal, C., & Kassahun, A. (2023). Plant disease detection using drones in precision agriculture. *Precision Agriculture*, 24, 1–20.

[12] Borhani, Y., Khoramdel, J., & Najafi, E. (2022). A deep learning based approach for automated plant disease classification using vision transformer. *Scientific Reports*, 12(1), 11554.

[13] Jackulin, C., & Murugavalli, S. (2022). A comprehensive review on detection of plant disease using machine learning and deep learning approaches. *Measurement: Sensors*, 24, 100441.

[14] Mohanty, S. P., Hughes, D. P., & Salathé, M. (2016). Using deep learning for image-based plant disease detection. *Frontiers in Plant Science*, 7, 1419.

[15] Rokhman, N., & Usuman, I. (2022, October). Systematic review of the early detection and classification of plant diseases using deep learning. In IOP Conference Series: Earth and Environmental Science (Vol. 1097, No. 1, p. 012042). IOP Publishing.

[16] Khan, R. U., Khan, K., Albattah, W., & Qamar, A. M. (2021). Image-based detection of plant diseases: From classical machine learning to deep learning journey. *Wireless Communications and Mobile Computing*, 2021, 1–13.

[17] Ashwinkumar, S., Rajagopal, S., Manimaran, V., & Jegajothi, B. (2022). Automated plant leaf disease detection and classification using optimal MobileNet based convolutional neural networks. *Materials Today: Proceedings*, 51, 480–487.

[18] Nayagam, M. G., Vijayalakshmi, B., Somasundaram, K., Mukunthan, M. A., Yogaraja, C. A., & Partheeban, P. (2023). Control of pests and diseases in plants using IOT Technology. *Measurement: Sensors*, 26, 100713.

[19] Ngongoma, M. S., Kabeya, M., & Moloi, K. (2023). A review of plant disease detection systems for farming applications. *Applied Sciences*, 13(10), 5982.

[20] Ampatzidis, Y., De Bellis, L., & Luvisi A. (2017). iPathology: Robotic applications and management of plants and plant diseases. *Sustainability* 9(6), 1010. https://doi.org/10.3390/su9061010

[21] Li, L., Zhang, S., & Wang, B. (2021). Plant disease detection and classification by deep learning – A review. *IEEE Access*, 9, 56683–56698.

[22] Ampatzidis, Y., De Bellis, L., & Luvisi, A. (2017). iPathology: Robotic applications and management of plants and plant diseases. *Sustainability*, 9(6), 1010.

[23] www.kaggle.com/datasets/vipoooool/new-plant-diseases-dataset

11 Apple Leaves Disease Detection Using Deep Learning

Smita Maurya, Arunima Jaiswal, Nitin Sachdeva, and Kanchan Sharma

11.1 INTRODUCTION

Agriculture in India is a fundamental source of food provision, playing a vital role in ensuring sustenance and nutritional security for its vast population through the cultivation of diverse crops and food commodities. In India, apples are not as widely grown as other fruits like mangoes or bananas; they play a significant role in the economy and agriculture of these regions. Apple trees are susceptible to various pests and diseases, which can cause significant damage if not managed properly. Common diseases affecting apple leaves include apple scab, powdery mildew, and fire blight. These diseases can reduce fruit quality and yield. Fungal diseases, such as apple scab, are prevalent in apple orchards. Apple scab affects leaves, fruit, and stems, leading to reduced fruit quality and economic losses. Insects like codling moths and aphids can damage apple leaves and fruit. Codling moths, in particular, can lead to fruit damage, reducing marketable yield. Finding plant diseases and pests quickly is a big problem for farmers. Detecting apple leaf diseases early is essential to making sure we can keep growing apples sustainably. Insects can harm crops and reduce their productivity. To address these challenges, apple growers in India are adopting modern agricultural practices, including the use of disease-resistant apple varieties, integrated pest management techniques, and improved irrigation systems.

In the past, plant disease detection primarily relied on manual visual inspections and basic laboratory tests. However, these methods had limitations, including subjectivity in interpretation, late detection of advanced diseases, limited accuracy in identifying subtle symptoms, time-consuming field surveys, and the need for resource-intensive laboratory analyses. These constraints hindered early intervention, accurate diagnosis, and efficient disease management, often leading to reduced crop yields and economic losses.

Modern plant disease detection employs advanced technologies like machine learning, remote sensing, and molecular diagnostics. Remote sensing techniques involve using drones and satellites to capture multispectral and hyperspectral imagery, enabling early disease detection by identifying subtle changes in plant health. Machine learning algorithms, particularly CNNs, analyze images to classify

DOI: 10.1201/9781003484608-11

diseases based on visual symptoms [1, 2]. These contemporary methods offer higher accuracy and early detection.

This study is founded on deep learning principles, employing Convolutional Neural Networks (CNNs) for apple leaf disease classification. The research encompasses a total of five models: four models: CNN,ResNet50, VGG16, and VGG19 and one hybrid model, CNN-KNN. The paper is divided into seven sections: The second section reviews previous research on apple leaf disease detection; the third section describes types of apple leaf diseases; the fourth section contains dataset description; the fifth section outlines the proposed methodology; the sixth section presents the paper's results; and the last section concludes the study.

11.2   RELATED WORK

Before, the only way to identify plant diseases was by looking at the leaves manually. People would inspect the leaves or use a book to find out which disease it might be [3]. But this method had three big problems: First, it was not very accurate. Second, it was impossible to check every single leaf. And finally, it took a lot of time.

As science and technology have improved, we now have better ways to identify plant diseases with greater accuracy [4]. There are two main approaches: image processing and deep learning. In image processing, different techniques like filtering, clustering, and studying image patterns are used to find the damaged areas and identify diseases [5]. In deep learning, we use neural networks to detect diseases. Yong Zhong et al. [6], the dataset comprising 2,462 images depicting six distinct apple leaf diseases served as the foundation for our data modeling and method evaluation.

Prakhar Bansal et al. [3] provided a manuscript that provides an overview of various approaches, including image processing, machine learning, and deep learning, employed for the detection of severe diseases in paddy plants. Through a comparative analysis of these methods, it became evident that deep learning techniques exhibited greater promise in the realm of paddy plant disease detection, surpassing the efficacy of the other two methodologies.

Bonkra et al. [7] conducted a scientometric analysis utilizing a sample of 214 documents focusing on the identification of apple leaf diseases. These documents were retrieved through a scientific search conducted on the Scopus database for the period spanning 2011 to 2022.

According to Amit Gawadel et al. [8], the proposed system is designed to detect various apple leaf diseases at an early stage, serving as an early warning mechanism for farmers and nearby research institutes. This timely detection enables them to initiate appropriate measures for disease control.

The study of Peng Jiang et al. [9] demonstrated the effectiveness of the innovative INAR-SSD model as a high-performance solution for early apple leaf disease diagnosis. Our model showcases superior capabilities in real-time disease detection, boasting higher accuracy and faster detection speeds when compared to previous methods.

According to Sangeetha K et al. [10], the primary objective of this proposed model is to streamline the complexity associated with apple leaf disease classification through deep learning techniques. This system has achieved an impressive validation accuracy of 93.3% on the apple leaf disease dataset, surpassing several existing state-of-the-art methods. With an average processing time of just 14 seconds per image, this system is not only efficient but also practical for adoption by farmers. It simplifies the apple leaf disease classification process, enabling early diagnosis and timely treatment of the disease, thereby benefiting agricultural practices.

Saraansh Baranwal et al. [11] provided a validation process involving an extensive dataset sourced from the Plant Village, encompassing diverse disease samples and healthy specimens. To ensure robust results, they employed a combination of image filtering, compression, and generation techniques to expand and fine-tune the training dataset. The trained model exhibited exceptional performance, achieving consistently high accuracy scores across all classes. The overall accuracy, computed over the entirety of the dataset, which includes 2,561 labeled images, stands at an impressive 98.54%.

Swati Singh et al. [12] provided experimental results that showcase the superiority of the proposed model, achieving an outstanding accuracy rate of 99.2%. This not only surpasses the performance of other CNN-based models but also outperforms the standard machine learning classifiers, underscoring the model's exceptional capability in disease classification.

Merugu Sai et al. [13] incorporated pre-trained CNN models to extract features from our dataset, and subsequently, applied these CNN models for classification. In our comparative analysis, achieved remarkable accuracies, with the standard CNN model surpassing the 93% accuracy threshold. Among the pre-trained models, the Inception V3 model delivered an accuracy of 92%, while the VGG16 and VGG19 models achieved 62% and 63% accuracy, respectively.

S. Alqethami et al. [14] provided a proposed method that not only distinguishes between healthy and diseased plant leaves but also goes a step further. It furnishes valuable recommendations for tailored solutions corresponding to each identified plant disease category, thereby offering actionable insights to combat these agricultural challenges.

The objective of the study by Kulbir Kaur Sandhu et al. [15] was to forecast various diseases affecting apple leaves, including but not limited to apple scab and marssonina. To achieve this, we employed a range of algorithms such as K nearest neighbor (KNN), support vector machine (SVM), classification decision tree, regression decision tree, and Naïve Bayes.

11.3 TYPES OF APPLE LEAVES DISEASES

There are several types of diseases that can affect apple tree leaves. These diseases can be caused by various pathogens, such as fungi, bacteria, and viruses. Here are some common types of apple leaf diseases:

11.3.1 APPLE SCAB (VENTURIA INAEQUALIS): Apple scab is one of the most common and damaging diseases affecting apple trees. It is caused by a fungus and appears as dark, scaly lesions on the leaves. Severe infections can lead to defoliation and reduced fruit quality.

11.3.2 CEDAR APPLE RUST (GYMNOSPORANGIUM JUNIPERI-VIRGINIANAE): Cedar apple rust is a fungal disease that affects both apple trees and cedar trees. It causes yellow-orange spots on apple leaves with raised lesions that may have a rusty appearance. This disease often requires both apple and cedar hosts to complete its life cycle.

11.3.3 APPLE POWDERY MILDEW (PODOSPHAERA SPP.): Powdery mildew is a fungal disease that forms a white, powdery coating on the surface of apple leaves. It can lead to distorted growth and reduced fruit quality if left uncontrolled.

11.3.4 APPLE BLACK SPOT (APPLE FROGEYE LEAF SPOT, BOTRYOSPHAERIA OBTUSA): Apple black spot is caused by a fungal pathogen and results in dark, circular lesions with a distinct black center on apple leaves. It can weaken the tree and reduce fruit yield.

11.3.5 APPLE RUST (PHRAGMIDIUM SPP.): Apple rust diseases are caused by various species of rust fungi. They produce yellow or orange pustules on the undersides of apple leaves and can lead to leaf drop if left untreated.

11.3.6 APPLE BLOTCH (PHYLLOSTICTA SPP.): Apple blotch is a fungal disease that causes irregularly shaped, dark lesions on apple leaves. In severe cases, it can lead to defoliation and reduced fruit quality.

11.3.7 APPLE LEAF SPOT (FABRAEA SPP.): Apple leaf spot diseases are caused by different species of fungi. They result in round, dark lesions on apple leaves and can reduce the tree's overall health.

11.3.8 FIRE BLIGHT (ERWINIA AMYLOVORA): While fire blight primarily affects blossoms, shoots, and branches, it can also cause wilting and browning of apple leaves. This bacterial disease can be very destructive to apple trees.

11.3.9 APPLE MOSAIC VIRUS: Apple mosaic virus is a viral disease that causes mottling, mosaics, and distortions in the leaves. Infected trees may also exhibit reduced growth and fruit quality.

11.3.10 APPLE LEAF CURLING MIDGE (DASINEURA MALI): This is caused by a tiny gall midge larva that feeds on apple leaves, causing them to curl and form blister-like galls.

Proper sanitation, cultural practices, and the use of fungicides or pesticides, when necessary, can help prevent and manage these apple leaf diseases. It is essential to identify the specific disease correctly to apply appropriate control measures effectively. Consult with a local agricultural extension office or a horticulturist for guidance on disease management in your specific area.

TABLE 11.1
Dataset Description

Name	No. Of Images	Scientific Name
Apple__Apple_scrab	753	Venturia inaequalis
Apple__Black_rot	800	Apple Frogeye Leaf Spot, Botryosphaeria obtusa
Apple__Cedar_Apple_rust	800	Phragmidium spp
Apple__healthy	1145	Malus domestica

11.4 DATASET DESCRIPTION

For this study, one dataset has been utilized. The set is called 'Plant Village Dataset'. The dataset has been sourced from a website known as Kaggle. The first set has 3366 pictures. As indicated in Table 11.1, the set has four groups called classes.

- The groups are

1. Apple__Apple_scrab
2. Apple__Black_rot
3. Apple__Cedar_Apple_rust
4. Apple__healthy

11.5 METHODOLOGY USED

Detecting apple leaf diseases using CNNs involves a specific methodology tailored to deep learning techniques, as mentioned in Figure 11.8. Here is a step-by-step methodology for apple leaf disease detection using CNNs:

11.5.1 DATA COLLECTION

- Compile a comprehensive dataset of apple leaf images encompassing both healthy leaves and leaves afflicted with various diseases. Ensure meticulous labeling of the dataset with corresponding disease classes. Sample images from the dataset are shown in Figure 11.1.

11.5.2 DATA PREPROCESSING

- Resize and standardize the images to a consistent size (e.g., 224x224 pixels) to ensure compatibility with the CNN architecture.
- Augment the dataset by applying random transformations like rotations, flips, and brightness adjustments to increase its diversity and improve the model's generalization.

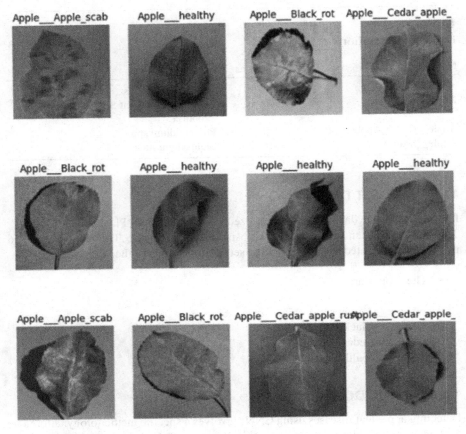

FIGURE 11.1 Sample image batch.

11.5.3 DATA SPLIT

- Divide the dataset into three subsets: training, validation, and testing. Common splits are 70% for training, 15% for validation, and 15% for testing. Ensure that each subset has a proportional representation of each class.

11.5.4 IMAGE CLASSIFICATION

- Four architectural models have been used in this study for the classification of potato leaf images into four categories Apple__Apple_scrab,Apple__Black_ rot, Apple__Cedar_Apple_rust and Apple__healthy
- These include CNN, ResNet50, VGG16,VGG19 and one hybrid model CNN-KNN.

11.5.5 Transfer Learning

- Utilize transfer learning by fine-tuning a pre-trained CNN model. This involves removing the last few layers of the network and adding new layers for classification.
- Freeze the early layers (feature extraction layers) to retain the knowledge learned from the pre-trained model.

11.5.6 Model Training

- Train the CNN model on the training dataset using an appropriate loss function (e.g., categorical cross-entropy) and an optimizer (e.g., Adam or Stochastic Gradient Descent (SGD). Images to train have to select as shown in Figure 11.2(a)(b).
- Continuously monitor the model's performance on the validation set to mitigate overfitting. Implement strategies such as early stopping and learning rate scheduling as safeguards.

11.5.7 Hyperparameter Tuning

- Experiment with different hyperparameters, such as learning rate, batch size, and network architecture, to optimize model performance.

11.5.8 Model Evaluation

- After training the model using the Train and Validation dataset, we assess the model's performance on the Test dataset by examining its accuracy and loss scores. Presenting this evaluation through two graphical representations: one illustrating the Training and Validation Accuracy as shown in Figure 11.7, and the other depicting the Training and Validation Loss of the model. Additionally, analyzing the model's performance using a confusion matrix, and we showcase a selection of sample predictions alongside their associated confidence scores. Confusion matrix of four different models are shown in Figures 11.3, 11.4, 11.5, and 11.6.
- Also calculated the Average Time per Epoch for each model, providing insights into computational efficiency. To further evaluate model performance, we generate a comprehensive classification report. This report includes key statistics such as precision, recall, F1-score, and support for each class as mentioned in Table 11.2, and it encompasses both macro and weighted averages. Additionally, the report provides an overall accuracy score, summarizing the model's classification process across all classes. Final sample prediction is depicted in Figure 11.10.

The whole process is like solving a puzzle, where each step helps figure out if there's a problem with the apple leaf.

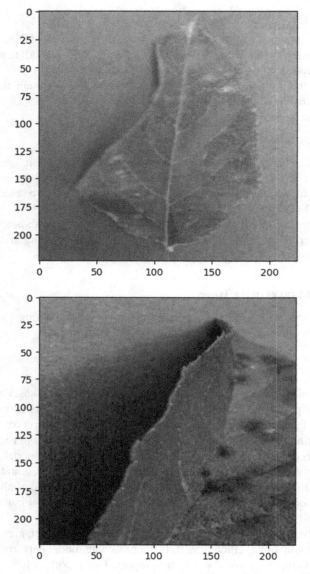

FIGURE 11.2   (a) First image to predict ResNet50 and (b) First image to predict VGG16.

11.6   RESULTS AND DISCUSSION

In the context of disease detection, five distinct models were employed for the task. A comprehensive comparison of these five base models revealed that VGG16 emerged as the top-performing model, demonstrating superior accuracy. The order of model performance, from least to most effective, is shown in Figure 11.9, is as

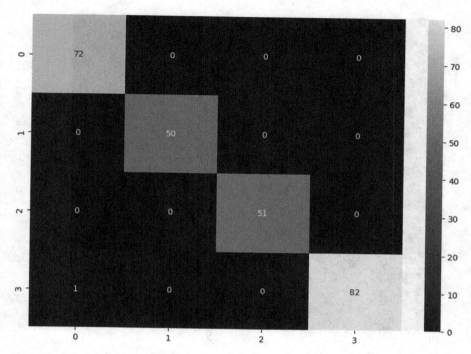

FIGURE 11.3 Confusions matrix (VGG16).

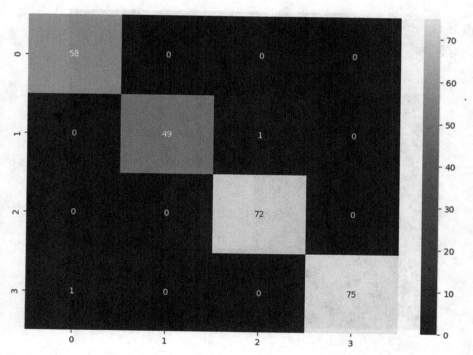

FIGURE 11.4 Confusions matrix (ResNet50).

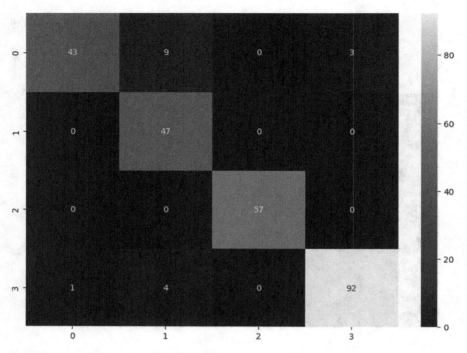

FIGURE 11.5   Confusions matrix (CNN).

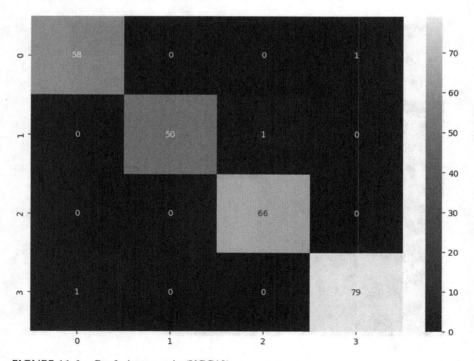

FIGURE 11.6   Confusions matrix (VGG19).

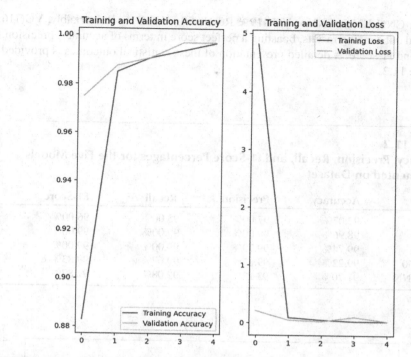

FIGURE 11.7 Fraining and validation accuracy & loss graphs.

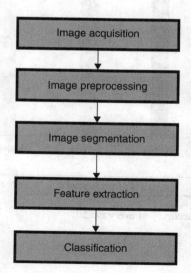

FIGURE 11.8 Methodology used.

follows: CNN-KNN<CNN < VGG19 < ResNet50 < VGG16. Remarkably, VGG16 achieved outstanding results, boasting a perfect score in terms of accuracy, precision, recall, and f1-score. A detailed presentation of these statistical outcomes is provided in Table 11.2.

TABLE 11.2
Accuracy, Precision, Recall, and f1-Score Percentages for the Five Models Implemented on Dataset

Model	Accuracy	Precision	Recall	F1-Score
CNN	92.07%	97.00%	95.00%	96.00%
VGG19	98.96%	99.00%	99.00%	99.00%
VGG16	99.74%	99.79%	99.00%	99.00%
ResNet50	99.22%	99.87%	99.67%	99.83%
CNN-KNN	91.70%	93.23%	92.08%	92.87%

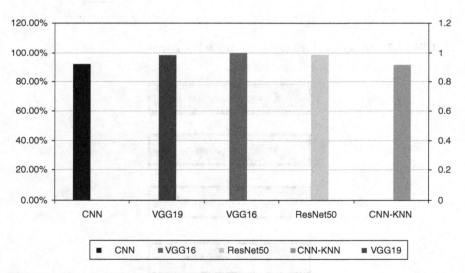

FIGURE 11.9   Graph of accuracy of every model.

FIGURE 11.10 Sample prediction with their confidences.

11.7 CONCLUSION

This study has successfully demonstrated the efficacy of deep learning techniques for the detection of diseases in apple leaves. This approach, which utilized CNNs and extensive image datasets, achieved promising results in terms of accuracy and efficiency. study on disease detection utilized five distinct models, and through a comprehensive comparison, VGG16 emerged as the clear frontrunner, showcasing remarkable superiority in terms of accuracy, as shown in Table 11.2. The performance hierarchy of these models, from least to most effective, is as follows: CNN-KNN<CNN < VGG19 < ResNet50 < VGG16. This paper has shown that deep learning can play a vital role in automating the detection and monitoring of diseases in apple trees, which is crucial for early intervention and crop management.

Looking forward, several exciting avenues await exploration in the realm of apple leaf disease detection using deep learning. Firstly, there is a pressing need to enhance the robustness of our models, accounting for variations in environmental conditions, lighting, and leaf orientations. The development of a multi-class classification system to detect various apple tree diseases simultaneously is another critical direction for future research. Additionally, exploring real-time monitoring capabilities through technologies like drones and remote sensing, coupled with deep learning, can revolutionize disease management practices in orchards. Advanced techniques such as data augmentation and transfer learning should be investigated further to make the most of limited labeled datasets.

REFERENCES

[1] Liu, Bin, Yun Zhang, DongJian He, Yuxiang Li, 2018. "Identification of Apple Leaf Diseases Based on Deep Convolutional Neural Networks," *Symmetry* 10(1): 11. https://doi.org/10.3390/sym10010011

[2] Chao, Xiaofei, Guoying Sun, Hongke Zhao, Min Li, Dongjian He, "Identification of Apple Tree Leaf Diseases Based on Deep Learning Models" *Symmetry* 12(7): 1065 (2020). https://doi.org/10.3390/sym12071065

[3] Prakhar Bansal, Rahul Kumar, Somesh Kumar, "Disease Detection in Apple Leaves Using Deep Convolutional Neural Network," *Agriculture* 11: 617 (2021).

[4] Jose V. Donu, K. Santhi, "Real Time Classification and Detection of Apple Leaf Diseases at Early Stage by Employing Enhanced Paddy Field Pattern Search Method (EPF-PSM) with 2LC," *Indian Journal of Science and Technology* 16: 1682–1693 (2023).

[5] S. Alqethami, B. Almtanni, W. Alzhrani, M. Alghamdi, "Disease Detection in Apple Leaves Using Image Processing Techniques", *Engineering Technology and Applied Science Research*, 12(2): 8335–8341 (Apr. 2022).

[6] Yong Zhong, Ming Zhao, "Research on Deep Learning in Apple Leaf Disease Recognition," *Computers Electronics in Agriculture* 168: 105146 (2020).

[7] Bonkra, Anupam, Pramod Kumar Bhatt, Joanna Rosak-Szyrocka, Kamalakanta Muduli, Ladislav Pilař, Amandeep Kaur, Nidhi Chahal, Arun Kumar Rana, "Apple Leave Disease Detection Using Collaborative ML/DL and Artificial Intelligence Methods: Scientometric Analysis," *International Journal of Environmental Research and Public Health* 20(4): 3222 (2023). https://doi.org/10.3390/ijerph20043222.

[8] Amit Gawade, Subodh Deolekar, Vaishali Patil, "Early-Stage Apple Leaf Disease Prediction Using Deep Learning," *Bioscience Biotechnology Research Communication* 14: 40–43 (2021).

[9] Peng Jiang, Yuehan Chen, Bin Liu, Dongjian He, Chunquan Liang, "Real-Time Detection of Apple Leaf Diseases Using Deep Learning Approach Based on Improved Convolutional Neural Networks," *IEEE* 7: 59069–59080 (2022). ISSN: 2169-3536.

[10] K Sangeetha, P Vishnu Raja , P Rima , M Pranesh Kumar , S Preethees, "Apple Leaf Disease Detection using Deep Learning," *IEEE*: 1063–1067 (2022). doi: 10.1109/ICCMC53470.2022.9753985

[11] Saraansh Baranwal, Siddhant Khandelwal, Anuja Arora, "Deep Learning Convolutional Neural Network for Apple Leaves Disease Detection (February 21, 2019)." Proceedings of International Conference on Sustainable Computing in Science, Technology and Management (SUSCOM), Amity University Rajasthan, Jaipur - India, February

26–28, 2019, SSRN. https://ssrn.com/abstract=3351641 or http://dx.doi.org/10.2139/ssrn.3351641

[12] Swati Singh, Isha Gupta, Sheifali Gupta, Deepika Koundal, Sultan Aljahdali, Shubham Mahajan,   Amit Kant Pandit, "Deep Learning Based Automated Detection of Diseases from Apple Leaf Images," *Computers, Materials & Continua* *71*(1): 1849–1866 (2022). https://doi.org/10.32604/cmc.2022.021875.

[13] Merugu Sai Charan, Mohammed Abrar, Bejjam Vasundhara Devi,"Apple Leaf Diseases Classification Using CNN with Transfer Learning,"*International Journal for Research in Applied Science and Engineering Technology* ISSN 10(VI): 1905-1912 (2022).

[14] S. Alqethami, B. Almtanni, W. Alzhrani, M. Alghamdi,"Disease Detection in Apple Leaves Using Image Processing Techniques," *Engineering, Technology and Applied Science Research* 12: 8335–8341 (2022).

[15] Kulbir Kaur Sandhu, "Apple Leaves Disease Detection Using Machine Learning Approach," *International Journal of Computer Science and Information Technology Research* ISSN 9(1): 127–135 (2021).

12 Comprehensive Approaches for a Recommendation System in Precision Agriculture

Ramesh Patra, Mamata Garanayak, and Bijay Kumar Paikaray

12.1 INTRODUCTION

It is never undervalued that for the majority of people, agriculture serves as their principal source of income. In our nation, agriculture has been practiced for thousands of years. As technology and equipment have advanced, traditional farming methods have been superseded. The need for sustainable practices, the growing global population, and climate change have made it imperative to find novel ways to increase agricultural output. With the help of cutting-edge technologies, precision agriculture has become a game-changing solution to these problems. The main role of precision agriculture is the integration of advanced technology like machine learning algorithms [1]. In the face of an ever-growing global population and the challenges posed by climate change, the imperative to enhance agricultural productivity has become more critical than ever. Precision agriculture, with its data-driven and technology-centric methodologies, offers a promising avenue to maximize agricultural yield while minimizing resource inputs. One of the pivotal components of precision agriculture is the development and implementation of advanced recommendation systems tailored to the unique needs of modern farming [2].

The contemporary agricultural landscape is characterized by a complex interplay of factors, ranging from climate variability and soil health to crop genetics and pest management. Traditional farming practices, although resilient, are proving insufficient in meeting the escalating demands for food production. This necessitates a paradigm shift toward precision agriculture, wherein data analytics, sensor technologies, and machine learning converge to optimize every facet of the farming process.

At the heart of precision agriculture lies the recommendation system, a sophisticated tool designed to provide farmers with actionable insights and guidance. This comprehensive approach aims not only to boost yield but also to ensure sustainability

DOI: 10.1201/9781003484608-12

by minimizing environmental impact and resource wastage. The following discussion delves into the multifaceted aspects of recommendation systems in precision agriculture, exploring their significance, challenges, and the diverse methodologies employed for their optimization [3].

The expansion of the agricultural sector is a big concern for both our nation's economy and farmers since it makes a significant contribution to both. Over 60% of Indian workers are employed in agriculture, which supports up to 25% of the country's Gross Domestic Product (GDP). Notwithstanding advancements in the field of agriculture, issues have persisted, aggravating Indian farmers and citizens. It is believed that poor growing, shipping, storage, and harvesting practices for government-subsidized crops cause one-fourth of India's produce to be lost. Farmers face several challenges, including unequal land distribution, a system of land tenure, holdings that are divided and fragmented, patterns of cropping, volatility and fluctuations, and labor conditions. Inadequate farming practices also lead to other issues that ultimately impact the nation's economy as a whole. The slowing of agricultural growth remains a critical issue for India's agricultural sector. One of the main causes of India's agriculture's slower growth is the decline in funding for agricultural research and development. As a result, certain technologies and approaches are required to collaborate and develop a system for farmers that addresses their issues [4].

Innovations in data analytics and technology have led to a revolutionary change in the agricultural landscape: precision agriculture. By utilizing real-time data, sensors, and advanced algorithms, this paradigm shift seeks to optimize agricultural techniques' efficiency. Optimizing agricultural productivity while minimizing resource inputs is at the core of this evolutionary process.

The emphasis on crop output prediction in agriculture provides a potentially useful approach to solving the current problem. In the absence of accurate harvest data, it is the process of projecting a crop's yield. Accurate and timely agricultural production projections are essential for defining future policies, planning imports and exports, and evaluating future hunger potential. This will also be very beneficial to the farmers, whose livelihoods are dependent on the success of their crops. The farmer may concentrate on implementing sustainable agriculture practices, such as judicious water use, fostering soil health, and lowering pollution levels, without worrying about crop output or returns from the field. All of these can be anticipated with a prior estimate of crop productivity [5]. Various types of classifiers, including Decision Tree, Logistic Regression, Naïve Bayes Classifier, and Random Forest, have been used in the analysis of crop prediction using machine learning.

12.2   BASIC CONCEPT

12.2.1   MACHINE LEARNING

These sensors record data, which is then stored on the microcontroller. Machine learning techniques, such as Random Forest, are then used to analyze the data and provide recommendations for the growth of a suitable crop. A machine learning algorithm is a collection of guidelines used to carry out tasks, usually in the form of predicting output values from a given set of input variables or finding new data insights and patterns. Machine learning (ML) enables learning through algorithms. ML algorithms can be classified into four categories: reinforcement, unsupervised, semi-supervised, and supervised. Every type and variation has a set of benefits based on your budget and the amount of speed and precision you demand. Various types of ML applications are available for agriculture prediction. The types of ML applications are Decision Trees, Logistic Regression, Naïve Bayes classifiers, and Random Forests. By using these methods, farmers can benefit from crop selection [6].

12.2.2 DECISION TREE

A decision tree is a type of learning technique used for predicting values or classes based on given data. It works by creating a tree structure where each node represents a decision based on data features, and the branches lead to different outcomes. This technique is particularly useful for making predictions in both regression (predicting numerical values) and classification (assigning categories). Decision tree classifiers are built using a greedy approach, meaning they make decisions based on the most significant feature at each step. The tree structure represents the features and the predicted outcomes. Decision trees are commonly used for creating models from historical data (training data) to make predictions about new data.

In a decision tree, there are two main types of nodes: decision nodes and leaves. Decision nodes represent tests on specific properties, while leaves represent the final outcomes or predictions. Every branch in the tree represents a possible response to a particular test case [7].

THE ALGORITHM FOR BUILDING A DECISION TREE IN MACHINE LEARNING

- Selecting the Best Feature:

Evaluate each feature in the dataset to find the one that best separates or categorizes the data. This is often done by measuring impurity using metrics like Gini impurity or entropy.

- Creating a Decision Node:

Create a decision node based on the selected feature. This decision node represents a test or condition related to the chosen feature.

- Splitting the Dataset:

Split the dataset into subsets based on the outcomes of the chosen feature. Each subset corresponds to a branch leading from the decision node.

- Recursion:

Repeat the process recursively for each subset. For each subset, select the best feature and create a decision node, splitting the data further. Continue this process until a stopping criterion is met.

• Stopping Criteria:
Define criteria to stop the tree-building process. This could include reaching a maximum depth, having a minimum number of samples in a node, or achieving perfect purity.

• Creating Leaf Nodes:
When the stopping criteria are met, create leaf nodes for the remaining subsets. The leaf nodes represent the final predictions or outcomes.

12.2.3   NAÏVE BAYES

Naïve Bayes classifiers are a bunch of classification methods rooted in Bayes' Theorem. Instead of just one method, it is like a family of related algorithms that share a common idea. The Naïve Bayes classifier stands out as a simple and effective way to create ML models that can predict quickly [8]. It comes in three main types:

12.2.3.1   Gaussian Naïve Bayes
Assumes data follows a normal distribution. It is good for continuous data and calculates probabilities based on how data is distributed.

12.2.3.2   Multinomial Naïve Bayes
Great for dealing with discrete data, especially in tasks like classifying text. It assumes data comes from a multinomial distribution, commonly used in natural language processing.

12.2.3.3   Bernoulli Naïve Bayes
Suitable for binary classification problems, where features are either present or absent. It works well when dealing with discrete data and assumes a Bernoulli distribution.

NAÏVE BAYES ALGORITHM:

Collect Data:
Gather a dataset where each observation is labeled with its corresponding class.

Preprocess Data:
Clean and preprocess the data as needed. This may include handling missing values, converting text data to numerical format, or scaling features.

Calculate Class Priors:
Calculate the prior probability of each class. This is the probability of each class occurring in the dataset without considering any features.

Calculate Feature Probabilities:
For Each Feature in the Dataset:

Calculate the likelihood of the feature given in each class. This involves estimating the probability distribution of each feature for each class.

Store Parameters:
Store the calculated probabilities for future predictions.

Prediction Phase:

Input Data:
Receive a new, unlabeled observation that needs to be classified.
Calculate Class Posteriors:

For Each Class:
Calculate the posterior probability of the class given the input data using Bayes' Theorem. This involves multiplying the prior probability of the class by the product of the likelihoods of the features given that class.

Predict Class:
Assign the class with the highest posterior probability as the predicted class for the input data.

12.2.4   LOGISTIC REGRESSION

Logistic Regression is a type of supervised ML used to predict probabilities in classification tasks. It examines the connection between a main outcome (dependent variable) and a factor (independent variable). It is a strong tool for decision-making [9].

Types of Logistic Regression:
• Binomial Logistic Regression:
Used when the outcome can be one of two possibilities, like Pass or Fail, represented by 0 or 1.

• Multinomial Logistic Regression:
Applied when the outcome can belong to one of several categories, such as "cat," "dog," or "sheep."

Ordinal Logistic Regression:
Useful when the outcome has three or more ordered categories, like "low," "medium," or "high."

Logistic Regression Algorithm:
• Collect Data:
Gather a labeled dataset where each observation has both features (independent variables) and the corresponding class labels (dependent variable).

• Preprocess Data:
Handle missing values, encode categorical variables, and scale/normalize features if necessary.

• Initialize Parameters:
Initialize the weights and bias (parameters) for the logistic regression model.

- Define the Sigmoid Function:

Define the sigmoid (logistic) function. The sigmoid function takes any real-valued number and transforms it to a value between 0 and 1.

$$S(z)=1+e-z1$$

where z is the linear combination of weights and features z=w0+w1x1+w2x2+...+ wnxn).

- Calculate Cost Function:

Define the cost function (also known as the logistic loss or cross-entropy loss) to measure the difference between predicted and actual class labels.

$$J(w)=-m1\Sigma i=1m[y(i)\log(y^\wedge(i)) +(1-y(i)) \log(1-y^\wedge(i))]$$

where m is the number of training examples, y(i) is the actual class label, and y^(i) is the predicted probability.

- Update Parameters:

Use optimization algorithms like gradient descent to minimize the cost function and update the weights and bias.

$$wj=wj-\alpha \partial wj \partial J(w)$$

where α is the learning rate.

- Repeat:

Repeat steps 4-6 until the cost converges to a minimum or reaches a predefined number of iterations.

Prediction Phase:

- Input New Data:

Receive new, unlabeled data that needs to be classified.

- Apply Trained Model:

Apply the learned weights and bias to calculate the probability using the sigmoid function.

$$y^\wedge=S(w0+w1x1+w2x2+...+wnxn)$$

- Thresholding:

Assign a class label based on a chosen threshold (commonly 0.5).

12.2.5   RANDOM FOREST

Random Forest is a supervised ML algorithm that is used for both regression and classification problems in ML. It solves a complex problem by combining multiple classifiers to improve the performance of this model. Random Forest is a classifier that averages the outcomes using many decision trees on various subsets of the given data to raise the dataset's projected accuracy.

An ensemble learning technique called a Random Forest combines the predictions from several decision trees to generate a prediction that is more reliable and accurate [10].

• Ensemble Learning:
An ML method called ensemble learning combines the predictions of several models to get a forecast that is more reliable and accurate. It is a method that enhances the learning system's overall performance by utilizing the combined intelligence of several models.

TYPES OF ENSEMBLE LEARNING

i. Bagging
With this approach, different models are trained using random subsets of the training set. Next, predictions from each of the separate models are integrated, usually through averaging.

ii. Boosting
Using this technique, a series of models are trained, with each new model concentrating on the mistakes produced by the prior model. Weighted voting is used to combine the predictions.

iii. Stacking
By utilizing this technique, features from one set of models' predictions are used as input for a different model. The second-level model makes the ultimate prediction.

WHY USE RANDOM FOREST

   i.   It takes less time than other algorithms.
   ii.   It gives high accuracy on the predicted outputs, and it runs smoothly even on large datasets.
   iii. When a significant amount of data is missing, it can nevertheless be accurate.

RANDOM FOREST ALGORITHM

• *Training Phase*
• *Collect Data*

Gather a labeled dataset with features and corresponding class labels.

• Randomly Sample Data:
Randomly select subsets (with replacement) from the dataset. Each subset is used to train a different decision tree.

- Build Decision Trees:

For Each Subset:

Choose a random subset of features at each split node during tree construction.

Create a decision tree using the selected features and a subset of the data.

Grow the tree until a predefined depth or a stopping criterion is met (e.g., minimum number of samples in a leaf node).

- Repeat:

Repeat steps 2-3 to build multiple decision trees. The randomness introduced by data and feature sampling enhances the diversity of the trees.

Prediction Phase:

- Input New Data:

Receive new, unlabeled data that needs to be classified.

- Aggregate Predictions:

For each decision tree:

Pass the new data through the tree to get a class prediction.

Collect predictions from all the trees.

- Majority Voting (Classification) or Average (Regression):

For classification: Assign the class label that receives the majority of votes from all trees.

For regression: Take the average of the predicted values.

12.3   RELATED WORKS

Burdett et al. [11] compare ML and statistical approaches to crop yield prediction in precision agriculture. It examines many techniques to examine the point-by-point relationship between soil properties and crop yields, such as Random Forests, Decision Trees, Multiple Linear Regression, And Artificial Neural Networks. In this case, the cross-validation technique is used to test the model and determine the characteristics of the soil and topography while making crop predictions. The study's dataset consists of georeferenced grid soil samples and yield measurements made with a full-size combine fitted with a GPS receiver and a commercial yield sensing device. The limits of ML techniques in identifying significant correlations between variables are acknowledged by the authors. The research was carried out in seventeen Southwestern Ontario fields that were under the management of a single cash cropping business.

Hakkim et al. [12] focus on site-specific management. A technology-based farm management approach called precision farming, sometimes called precision agriculture, strives to maximize output, profitability, sustainability, and preservation of land resources. Achieving the best results entails identifying, analyzing, and managing temporal and spatial variability within fields using information and technology. It highlights the difficulties presented by market-based international rivalry in agriculture as well as the growing reliance on irrigation plans for the world's food supply. It talks about using technology to organize and monitor agricultural operations, like vaginal thermometers and sensors on animal collars. The utilization of inputs based on the appropriate amount, timing, and location, a practice known as "Site-Specific Management", is highlighted in the study. The use of technology for planning and monitoring agricultural operations is also mentioned in the report. Examples of this technology include sensors on animal collars and vaginal thermometers. It refers to the internet's capacity to provide information on new technologies for agricultural production, including details on upcoming goods, technical specs, and software updates.

Bondre et al. [13] discuss various works related to ML in crop prediction and recommendation. The ML system is broken down into three parts sampling, backpropagation algorithm, and weight updating. This research suggests an information-driven method for predicting soil moisture utilizing AI tools such as Relevance and Support Vector Machines. To gather information on soil moisture and build a prediction system that is particular to a given location, they design a responsive remote sensor hub.

Garanayak et al. [14] begin by identifying the problem of poor agricultural conditions and the impact of temperature variations on crop production. The authors then suggest developing a recommendation system for agricultural yield using ML regression techniques, such as Random Forest, Linear Regression, Decision Tree Regression, Polynomial Regression, And Support Vector Regression. The meteorological data, including area and production, is used as input for the regression models. The paper compares the performance of different regression methods and calculates the overall percentage improvement over existing methods, which is found to be 3.6%. The authors also mention the preprocessing steps involved, such as handling missing values and transforming data using techniques like IsNull() and label encoder().

The paper focuses on five different crops (rice, ragi, potato, gram, and onion) and aims to accurately predict their production forecast. The paper evaluates the performance of various ML methods, including Random Forest, Linear Regression, Decision Tree Regression, Polynomial Regression, and Support Vector Regression, for crop yield recommendation. The overall percentage improvement over existing methods achieved by the recommendation system is reported to be 3.6%.

Mahalakshmi et al. [15] suggest using ML techniques, including Adaptive Network Based Fuzzy Inference System (ANFIS), Support Vector Regression, and Linear Regression to predict future crop yields and offer precise crop recommendations; the agricultural dataset is analyzed using data mining tools, which produce crop suggestions depending on the particular needs and circumstances. To enhance the performance of the suggested model, data configuration and preparation are carried out, including the removal of less dependent and irrelevant characteristics from the dataset. The methodology proposed in this study makes use of ML techniques to offer precise crop recommendations based on a number of factors, including productivity, soil type, season, and availability of water. It is suggested that image processing might be used to identify crop diseases, which could improve the model's accuracy even more.

Singh et al. [16] do an experimental analysis to show how well the ML technique may be used to increase crop production productivity. After removing null values from the dataset, the authors sort it according to the selected criteria. The authors anticipate crop yield using a Support Vector Machine (SVM) approach. The efficacy of the suggested approach is assessed through the utilization of measures like F-measure, precision, accuracy, recall, and classification error. The proposed method predicts crop yields with a promising accuracy that can help with strategic planning for agricultural commodities and increase farmer revenues. The study emphasizes how crucial it is for agricultural research to use technology and ML to advance economic development. The focus is on using strong algorithms and massive datasets to uncover the functional relationship between yield and other interacting factors. F-measure, precision, accuracy, recall, and classification error are among the metrics used to evaluate the efficacy of the recommended approach.

Jaiswal et al. [17] use the KNN (K-Nearest Neighbors) method in the study for query analysis and optimization using Collaborative Filtering. The study emphasizes how crucial it is to alleviate farmer hardship and how successful it must be for government programs to be implemented to enhance India's agriculture industry. Through an online application, the system creates a profile of the fundamental requirements and suggests government programs to assist farmers based on their inquiries and requirements. To get the best results, the KNN algorithm is optimized by choosing the best value for k. Additionally, the system notifies farmers regularly about new government initiatives and programs as well as current agricultural trends.

Richa Kumari Karn et al. [18] initially focus on the difficulties farmers have in meeting the growing demand for food crops in regions with changing temperatures and expanding populations. Along with examining how environmental factors affect agricultural productivity, the research establishes important benchmarks for future regulation and oversight. The suggested model is put to the test and verified by comparing it to the most advanced yield prediction techniques.

The goal of the article is to increase the precision and promptness of wheat crop yield prediction, enabling researchers and farmers to assess risks and take preventive measures to preserve steady yields. The suggested model can help farmers, policymakers, and planners make informed decisions and take corrective and preventive action to increase agricultural yields. It was evaluated and validated through comparative analysis with state-of-the-art methodologies for yield prediction. To estimate crop yield, the study carried out numerous tests on test data utilizing different hyperparameter solutions that were derived using KNNDT.

Medar et al. [19] propose that the study increases agricultural yield rates and addresses issues that farmers encounter by utilizing ML techniques for crop yield prediction in the agriculture industry. The methodology of this paper is to gather and handle the datasets, eliminating any contaminants and, if required, standardizing the information. Store the data in databases after converting it to a format that can be supported. Utilize the necessary techniques for predicting crop yield. Acquire the ultimate outcomes. This research aims to apply ML techniques to crop selection to solve farmer and agriculture difficulties and increase crop yield rate, which will boost the Indian economy. The main advantage of this paper is it does not require the consideration of data model structures, potentially simplifying the prediction models. The disadvantages of this paper are less accuracy in terms of performance and less prediction of crops.

Maheswari et al. [20] propose to do a comparative analysis of ML algorithms for agricultural crop yield prediction. The goal of the study is to demonstrate how crucial feature selection and ML algorithm hyperparameter tuning are to crop production prediction. When compared to other models, the Random Forest algorithm is proven to have good accuracy. When it comes to crop yield prediction, the Random Forest method outperforms other models with a high degree of accuracy. It is stressed how crucial feature selection and hyperparameter adjustment are for these ML methods. Additionally, the XGBoost technique is used, which makes use of complex algorithms to increase prediction speed and accuracy for crop yield. When the Random Forest (RF) method is further used, crop prediction accuracy increases even more. The XGBoost algorithm is used, and its high accuracy of 99.31% in crop yield prediction offers encouraging results. The RF method increases crop prediction accuracy to 99.54%. Crop production prediction using ML techniques may be able to help with the urgent problem of agricultural productivity.

Prabhu et al. [21] aim to survey ML applications in precision agriculture. In order to boost productivity and crop yields, it seeks to investigate how ML techniques might be used to enable precise application of information sources, such as toxicants, seeds, manure, and effluent, at the exact moment of harvest. The process might also entail compiling and evaluating the results of the data gathered to offer perspectives on the advantages and difficulties of applying ML to precision farming. Case studies or other examples that highlight how ML may be used to precisely apply information sources, boost productivity, and increase crop yields could be included in the article. In light of the soil's properties and other abiotic factors, the paper's conclusion recommends using a yield recommendation apparatus to assist farmers in choosing the right crop to plant on their properties. Higher crop yields and productivity may result from this.

Devikarani et al. [22] examine a number of ML techniques, including K-NN, SVM, RF, and Neural Network (NN), which are used to analyze environmental conditions and provide recommendations for agricultural yield, crop protection, and tailored farmer support. Additionally, based on the farm's geolocation, the study addresses the usage of a predictor component that helps determine the best growth circumstances for particular plants by projecting rainfall for a year. The paper's goal is to present a thorough examination and critique of ML algorithms used in smart agriculture, particularly for farm management and crop growth prediction. It evaluates the performance of these algorithms using a range of criteria, including precision, recall, and F-score, offering insights into how well they assess environmental parameters and propose crop production. The goal of the study is to offer the best possible answer for smart agriculture while highlighting the need for more research in this area.

Kale et al. [23] aim to create an NN model for ML to forecast crop production and success rate using the Indian government's dataset. The goal is to provide yield projections for various crops so that farmers can make well-informed decisions about which crops to produce. The study utilized a dataset of two lakh 40 thousand records from 1998 to 2014 that was sourced from an Indian government website. A three-layer multilayer perceptron NN is used to create the model. The model makes use of both forward and backward propagation techniques and the Rectified Linear activation unit (ReLU) as the activation function. When utilizing the RMSprop optimizer, the model's starting accuracy of 45% is increased to 90% by adding more layers, modifying the weight and bias, and switching to the Adam optimizer. The study examines how variables depend on one another and uses linear correlation equations to shed light on the link between variables.

Mishra et al. [24] highlight the need for objective, statistically sound forecasts as well as the significance of timely and accurate crop production projections for policy decisions. The study offers insights into the use of ML methods in agriculture and how they might enhance crop production analysis and forecasts. In the field of agriculture, the authors assess novel approaches like Artificial Neural Networks, k-Means Clustering, K-Nearest Neighbor, Support Vector Machines, Decision Trees, Regression Analysis, Bayesian Belief Networks, Time Series Analysis, Artificial Neural Networks, and Information Fuzzy networks. Data from various locations are used to evaluate the second-order Markov chain model's performance in terms of temporal depth and dimensional coverage. This work aims to review the literature on the applicability of ML methods in agricultural crop production.

Vardhan et al. [25] aim to investigate precision agriculture, a novel farming technique that leverages ML to offer farmers advice on which crops to plant depending on site-specific parameters. The authors discuss data analysis and prediction using ML techniques, namely the RF algorithm for yield projection and the Naïve Bayes approach for crop classification. In order to prepare the dataset for training the ML models, the study also discusses data preprocessing, which includes eliminating noisy and missing values. The possible use of these ML models in a web application for in-depth study of real-world data is mentioned by the authors. The study emphasizes the utilization of historical data from farming activities to increase crop yields and recommend particular crops for planting.

Adamides et al. [26] utilize the review process used in this work based on the three-phase approach described in Kitchenham and Charters' 2007 review process: preparation, implementation, and reporting. The reporting phase is covered in the "Literature review findings" section, whereas the planning and execution phases are detailed in the paper. After conducting a quantitative analysis of the literature, the authors examined 27 different studies carried out over the previous 21 years involving human-robot collaboration (HRC) systems in agriculture. Safety, ergonomics, awareness, and productivity in HRC were among the significant and cutting-edge subjects covered in the review. In order to improve performance and put HRC in agriculture into practice, this paper reviews and offers a roadmap for HRC systems in agricultural jobs with an emphasis on safety, ergonomics, awareness, and productivity. It highlights novel and crucial HRC subjects like productivity, safety, ergonomics, and awareness. Together with the characteristics of HRC systems, performance metrics, and experimentation techniques, it also offers a roadmap for future research and development paths to enhance performance and apply HRC in agriculture.

M Keerthana et al. [27] focus on predicting crop diseases and raising agricultural output through the use of ensemble approaches and data mining tools. To create the ensemble model, the researchers employ a variety of classifiers, including K-Nearest Neighbor, RFs, Decision Trees, Neural Networks, and Naïve Bayes Classifiers. To get the data ready for analysis, they use data preprocessing methods, including transformation, feature selection, normalization, and cleaning. To improve performance, the ensemble model integrates many classifiers using the chi-square approach for feature selection. Model accuracy is assessed using F1 Score, precision, and recall. The findings demonstrate that weak classifiers such as KNN and SVM perform better when used in an ensemble setting, with an accuracy of 36 and 33, respectively. This paper aims to forecast crop loss caused by pest insects and create an ensemble model using data mining techniques to increase prediction accuracy. The goal is to assist farmers in taking the necessary actions to lessen the issue and improve production results. The outcomes demonstrate that weak classifiers such as KNN and SVM perform better when used in an ensemble setting, with an accuracy of 36 and 33, respectively. In order to prepare the data for analysis, the study also emphasizes the significance of data preprocessing methods such as transformation, feature selection, normalization, and cleaning.

12.4   OBJECTIVES

In agriculture, crop yield prediction is essential since it guides strategy and policy for feeding the world's expanding population. The weather is an uncontrollable factor that affects agricultural productivity. The survival of many crops is threatened by the erratic rise in temperature brought on by global warming. One important and commonly consumed cereal, wheat, is severely harmed by high temperatures. Furthermore, via ineffective crop management, farmers' attempts to satisfy the growing demand for grains are reducing the natural resources and nitrogen pool in the soil. To more thoroughly examine the relationship between yield and weather characteristics, the entire crop season must be divided into meaningful intervals due to the bias induced by the weather distribution pattern throughout the crop season. Predicting agriculture yields using Random Forest gives the best accuracy for the prediction of crops.
The following are the various objectives of this chapter:

1. To study and analyze ML algorithms for the prediction of crops.
2. The prediction model aims to predict crop yield based on various factors such as rainfall, fertilizer, soil color, pH, nitrogen, phosphorous, and potassium.
3. The prediction of the crop will increase efficiency using the RF algorithm.
4. The objective of predicting agriculture is to enhance the economic viability of farming operations.
5. Crop prediction helps farmers anticipate the quantity and quality of their produce, enabling better market planning.

12.5   PROBLEM STATEMENT

The problem statement in this paper typically revolves around helping farmers make informed decisions about which crops to cultivate based on various factors.

The primary goal is to optimize crop yield, considering factors such as soil type, weather conditions, and other environmental conditions. The system aims to provide personalized recommendations to farmers, taking into account the unique characteristics of their farmland.

In this paper, we have the various ML models (Decision Tree, Logistic regression, Naïve Bayes, and RF) and compare among various models we get the highest accuracy on RF; we take that model and use it to predict the crop.

THE MAJOR KEY POINTS OF THE PROBLEM STATEMENT ARE:

1. *Data Collection and Integration:* Gather relevant data such as soil characteristics, climate conditions, and historical crop yields.
2. *Feature Selection and Analysis:* Identify key features that influence crop growth and yield. Analyze the relationships between these features and historical crop performance.
3. *Model Development:* Build ML models that can predict suitable crops based on the input features. Consider using algorithms that can handle multi-variate data and adapt to different farming scenarios.
4. *User Interface:* Develop a user-friendly interface for farmers to input their data and receive personalized crop recommendations. The interface should be accessible and easy to use in diverse agricultural settings.
5. *Scalability and Adaptability:* Design the system to scale across various regions and adapt to different types of crops and farming practices.
6. *Accuracy and Validation:* Ensure that the recommendation system is accurate and reliable by validating it with real-world data. Continuously update and improve the system based on new information.
7. *Cost-effectiveness:* Consider the economic aspects of crop recommendations, aiming for a system that provides cost-effective solutions for farmers.

12.6   METHODOLOGY

The methodology in crop prediction involves a systematic approach to collecting, analyzing, and interpreting data to predict crop yields. We have drawn a flow chart of the methodology part as in Figure 12.1.

12.6.1   FLOW CHART OF THE METHODOLOGY

Figure 12.1 illustrates the comprehensive procedure from initial data collection and preprocessing to the application of various ML techniques, comparing their accuracies to choose the optimal model, culminating in the prediction of the crop by utilizing the chosen best model.

12.6.2   DATA COLLECTION

Data Collection is a difficult component for researchers because it is difficult to collect the data. It allows researchers to support investigations. In this work, data is

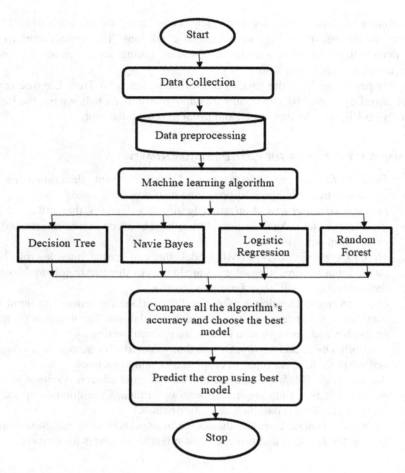

FIGURE 12.1 Proposed method.

collected from the Kaggle website and datasets contain 4513 train data and 17 types of crops for prediction as in Figure 12.2.

After the collection of dataset, the features such as soil color, Crop and Fertilizer are encoded as 1, 2, 3… as in Figure 12.3.

12.6.3 DATA PREPROCESSING

1. Drop the unused column from the dataset.

We have dropped unused column from the data set for preprocessing the data. After dropping one column the dataset is shown in Figure 12.4.

Then we checked for the missing values of the dataset. If there is some missing data then we cannot use in the prediction model. In the dataset there are no missing values as in Figure 12.5.

	District_Name	Soil_color	Nitrogen	Phosphorus	Potassium	pH	Rainfall	Temperature	Crop	Fertilizer	Link
0	Kolhapur	Black	75	50	100	6.5	1000	20	Sugarcane	Urea	https://youtu.be/2t5Am0xLTOo
1	Kolhapur	Black	80	50	100	6.5	1000	20	Sugarcane	Urea	https://youtu.be/2t5Am0xLTOo
2	Kolhapur	Black	85	50	100	6.5	1000	20	Sugarcane	Urea	https://youtu.be/2t5Am0xLTOo
3	Kolhapur	Black	90	50	100	6.5	1000	20	Sugarcane	Urea	https://youtu.be/2t5Am0xLTOo
4	Kolhapur	Black	95	50	100	6.5	1000	20	Sugarcane	Urea	https://youtu.be/2t5Am0xLTOo
...	...	...	...	...	...	...	...	...	...	...	...
4508	Pune	Black	130	80	150	7.0	1400	30	Sugarcane	MOP	https://youtu.be/2t5Am0xLTOo
4509	Pune	Black	135	80	150	7.0	1400	30	Sugarcane	MOP	https://youtu.be/2t5Am0xLTOo
4510	Pune	Black	140	80	150	7.0	1400	30	Sugarcane	MOP	https://youtu.be/2t5Am0xLTOo
4511	Pune	Black	145	80	150	7.0	1400	30	Sugarcane	MOP	https://youtu.be/2t5Am0xLTOo
4512	Pune	Black	150	80	150	7.0	1400	30	Sugarcane	MOP	https://youtu.be/2t5Am0xLTOo

4513 rows × 11 columns

FIGURE 12.2   Train Data.

	Soil_color	Nitrogen	Phosphorus	Potassium	pH	Rainfall	Temperature	Crop	Fertilizer
0	1	75	50	100	6.5	1000	20	1	1
1	1	80	50	100	6.5	1000	20	1	1
2	1	85	50	100	6.5	1000	20	1	1
3	1	90	50	100	6.5	1000	20	1	1
4	1	95	50	100	6.5	1000	20	1	1
...	...	...	...	...	...	...	...	...	...
4508	1	130	80	150	7.0	1400	30	1	3
4509	1	135	80	150	7.0	1400	30	1	3
4510	1	140	80	150	7.0	1400	30	1	3
4511	1	145	80	150	7.0	1400	30	1	3
4512	1	150	80	150	7.0	1400	30	1	3

4513 rows × 9 columns

FIGURE 12.3   Encoded Data.

	Soil_color	Nitrogen	Phosphorus	Potassium	pH	Rainfall	Temperature	Crop	Fertilizer
0	Black	75	50	100	6.5	1000	20	Sugarcane	Urea
1	Black	80	50	100	6.5	1000	20	Sugarcane	Urea
2	Black	85	50	100	6.5	1000	20	Sugarcane	Urea
3	Black	90	50	100	6.5	1000	20	Sugarcane	Urea
4	Black	95	50	100	6.5	1000	20	Sugarcane	Urea
...	...	...	...	...	...	...	...	...	...
4508	Black	130	80	150	7.0	1400	30	Sugarcane	MOP
4509	Black	135	80	150	7.0	1400	30	Sugarcane	MOP
4510	Black	140	80	150	7.0	1400	30	Sugarcane	MOP
4511	Black	145	80	150	7.0	1400	30	Sugarcane	MOP
4512	Black	150	80	150	7.0	1400	30	Sugarcane	MOP

4513 rows × 9 columns

FIGURE 12.4   Preprocessed Data.

```
dataset.isnull().sum()

District_Name      0
Soil_color         0
Nitrogen           0
Phosphorus         0
Potassium          0
pH                 0
Rainfall           0
Temperature        0
Crop               0
Fertilizer         0
Link               0
dtype: int64
```

FIGURE 12.5   Checking of missing value.

```
Crop
Sugarcane      1010
Wheat           859
Cotton          650
Jowar           394
Maize           350
Rice            309
Groundnut       177
Tur             126
Ginger          125
Grapes          125
Urad             99
Moong            99
Gram             78
Turmeric         55
Soybean          45
Masoor           12
Name: count, dtype: int64
```

FIGURE 12.6   Counting of different types of crops.

We have counted the different types of crops, fertilizer and soil type as shown in Figures 12.6, 12.7 and 12.8.

From Figure 12.6, it is evident that there are 16 distinct types of crops, and subsequent analysis reveals that sugarcane exhibits the highest count at 1010 instances, while Masoor is observed to have the lowest count at 12 instances.

In Figure 12.7, we conducted a tally of several fertilizer types, revealing a total of 18 distinct varieties. Subsequently, we meticulously counted the total no. of fertilizers across these diverse types for comprehensive analysis.

Figure 12.8 shows that there are six categories of several soils present in our dataset.

12.6.4   VISUALIZATION OF DATA

Visualization in research refers to the creation and use of visual representations to communicate, analyze, and interpret data. This can include various types of graphical

```
Fertilizer
Urea                        1364
DAP                          667
MOP                          571
19:19:19 NPK                 480
SSP                          417
Magnesium Sulphate           215
10:26:26 NPK                 156
50:26:26 NPK                 124
Chilated Micronutrient       108
12:32:16 NPK                 106
Ferrous Sulphate              68
13:32:26 NPK                  66
Ammonium Sulphate             50
10:10:10 NPK                  50
Hydrated Lime                 25
White Potash                  19
20:20:20 NPK                  15
18:46:00 NPK                   6
Sulphur                        6
Name: count, dtype: int64
```

FIGURE 12.7   Counting of different types of fertilizer.

```
Soil_color
Black          2260
Red            1224
Dark Brown      659
Reddish Brown   265
Light Brown      54
Medium Brown     51
Name: count, dtype: int64
```

FIGURE 12.8   Counting of different types of soil color.

elements such as charts, graphs, diagrams, maps, and other visual aids. The goal of visualization in research is to enhance the understanding of complex data sets, patterns, and relationships, making it easier for researchers, stakeholders, and audiences to grasp the information.

We have compared the temperature with rainfall as result we got that if the rainfall is between 300 to 1700 then there is a chance of increase or lose the temperature as in Figure 12.9.

The comparison of district name with temperature is shown in Figure 12.10 to found that different district has different levels of temperature.

The comparison of the Soil Color with temperature is shown in Figure 12.11 to find the average of different temperature.

Figure 12.12 shows a pie chart of the percentage of different type of crops. Sugarcane is the highest percentage which is giving 22.33%.

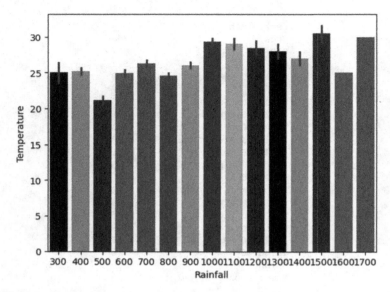

FIGURE 12.9 Comparison of temperature and rainfall.

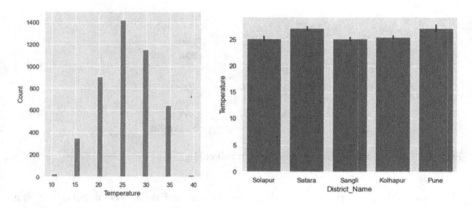

FIGURE 12.10 Comparison of district and temperature.

12.6.5 MACHINE LEARNING ALGORITHMS

We have used four models in this paper to extract the better one based on the best accuracy and used it to predict the crop.

The predictive models are:

12.6.5.1 Decision Tree

We extract the accuracy of the Decision Tree, which gives 79% accuracy, as in Figure 12.13.

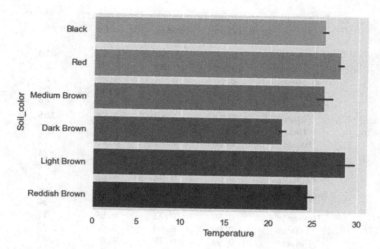

FIGURE 12.11 Comparison of soil color and temperature.

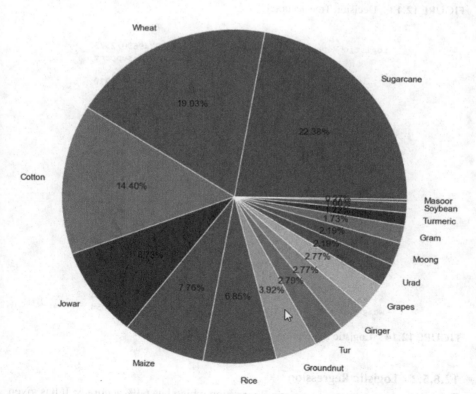

FIGURE 12.12 Percentage of different types of crops.

```
DecisionTree's Accuracy is : 0.7895125553914328
              precision    recall  f1-score   support

      Cotton       0.70      1.00      0.83       210
      Ginger       1.00      0.84      0.91        31
        Gram       0.00      0.00      0.00        25
      Grapes       1.00      1.00      1.00        29
   Groundnut       0.63      0.29      0.40        58
       Jowar       0.44      0.88      0.59       105
       Maize       1.00      0.88      0.94       109
      Masoor       0.00      0.00      0.00         6
       Moong       0.00      0.00      0.00        21
        Rice       0.73      0.60      0.66        96
     Soybean       0.00      0.00      0.00        14
   Sugarcane       1.00      0.95      0.98       308
         Tur       0.54      1.00      0.70        30
    Turmeric       0.50      0.71      0.59        17
        Urad       0.00      0.00      0.00        32
       Wheat       0.94      0.78      0.85       263

    accuracy                           0.79      1354
   macro avg       0.53      0.56      0.53      1354
weighted avg       0.78      0.79      0.77      1354
```

FIGURE 12.13   Decision Tree accuracy.

```
Logistic Regression's Accuracy is:  0.6403249630723782
              precision    recall  f1-score   support

      Cotton       0.56      0.49      0.52       210
      Ginger       0.64      0.90      0.75        31
        Gram       0.00      0.00      0.00        25
      Grapes       0.94      1.00      0.97        29
   Groundnut       0.45      0.78      0.57        58
       Jowar       0.53      0.60      0.57       105
       Maize       0.77      0.87      0.82       109
      Masoor       0.00      0.00      0.00         6
       Moong       1.00      0.19      0.32        21
        Rice       0.62      0.74      0.68        96
     Soybean       0.00      0.00      0.00        14
   Sugarcane       0.78      0.81      0.80       308
         Tur       0.22      0.07      0.10        30
    Turmeric       1.00      0.59      0.74        17
        Urad       0.00      0.00      0.00        32
       Wheat       0.56      0.64      0.60       263

    accuracy                           0.64      1354
   macro avg       0.50      0.48      0.46      1354
weighted avg       0.61      0.64      0.61      1354
```

FIGURE 12.14   Logistic Regression accuracy.

12.6.5.2   Logistic Regression

We extract the accuracy of Logistic Regression which has 64% accuracy. It has given the lowest accuracy as in Figure 12.14.

12.6.5.3   Naïve Bayes Classifier

The accuracy of the Naïve Bayes classifier was found to be 83%, as in Figure 12.15.

12.6.5.4   Random Forest

The accuracy of random forest was found 99% as in Figure 12.16.

```
NaiveBayes's accuracy is : 0.8316100443131462
              precision    recall  f1-score   support

      Cotton       0.87      0.88      0.87       210
      Ginger       1.00      1.00      1.00        31
        Gram       0.25      1.00      0.40        25
      Grapes       1.00      1.00      1.00        29
   Groundnut       0.69      0.34      0.46        58
       Jowar       0.63      0.42      0.50       105
       Maize       0.95      0.99      0.97       109
      Masoor       0.86      1.00      0.92         6
       Moong       0.83      0.24      0.37        21
        Rice       0.93      0.59      0.73        96
     Soybean       0.50      1.00      0.67        14
   Sugarcane       0.97      1.00      0.99       308
         Tur       0.78      0.83      0.81        30
    Turmeric       1.00      0.82      0.90        17
        Urad       0.38      0.19      0.25        32
       Wheat       0.87      0.95      0.91       263

    accuracy                          0.83      1354
   macro avg       0.78      0.77      0.73      1354
weighted avg       0.86      0.83      0.83      1354
```

FIGURE 12.15   Naïve Bayes accuracy.

```
RF's accuracy is : 0.9955301329394387
              precision    recall  f1-score   support

      Cotton       1.00      1.00      1.00       210
      Ginger       1.00      1.00      1.00        31
        Gram       1.00      1.00      1.00        25
      Grapes       1.00      1.00      1.00        29
   Groundnut       0.98      0.97      0.97        58
       Jowar       0.96      1.00      0.98       105
       Maize       0.99      0.99      0.99       109
      Masoor       1.00      1.00      1.00         6
       Moong       1.00      1.00      1.00        21
        Rice       1.00      1.00      1.00        96
     Soybean       1.00      1.00      1.00        14
   Sugarcane       1.00      1.00      1.00       308
         Tur       1.00      1.00      1.00        30
    Turmeric       1.00      1.00      1.00        17
        Urad       1.00      0.91      0.95        32
       Wheat       1.00      1.00      1.00       263

    accuracy                          0.99      1354
   macro avg       1.00      0.99      0.99      1354
weighted avg       0.99      0.99      0.99      1354
```

FIGURE 12.16   Random Forest accuracy.

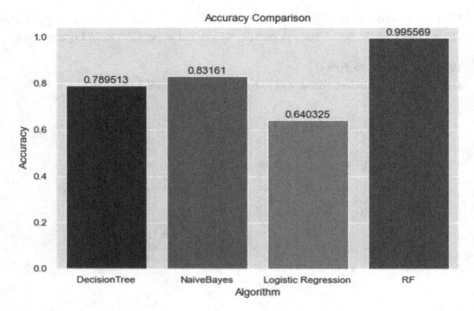

FIGURE 12.17 Comparison of models.

```
from sklearn.metrics import accracy_score
accuracy=accuracy_score(predicted_values, Ytest)
print('Random Forest Accuracy Score:{0:0.4f}'.format(Ytest,accuracy_score(predicted_values)))
```
Random Forest Accuracy Score: 0.9955

FIGURE 12.18 Random Forest accuracy prediction.

12.6.5.5 Compare all the Algorithms and Choose the Best One

In this section we compare the accuracy among the four models and choose that the best model is RF as it exhibits an accuracy of 99%, we compare the accuracy among the four models and choose the best model is RF, as it exhibits an accuracy of 99%, as in Figure 12.17.

12.6.5.6 Predict the Crop using the Best Model

The RF model has the highest accuracy; it is giving 99% as in Figure 12.18. So, we took that model to predict the crop.

Figure 12.19 illustrates the predictive process of crop selection by inputting diverse parameters, including soil color, nitrogen, potassium, phosphorus, pH, rainfall, temperature, and fertilizer quantity. Utilizing our RF model, the analysis identifies wheat as the optimal crop for cultivation based on these parameters.

```
Soil_color =4
Nitrogen =  115
Potassium = 50
Phosphorus = 40
pH = 6.5
Rainfall = 700
Temperature = 15
Fertilizer = 4
predict = recommendation(Soil_color,Nitrogen,Potassium,Phosphorus,pH,Rainfall,Temperature,Fertilizer)

crop_dict ={ 1: "Sugarcane", 2: "Wheat", 3: "Cotton", 4: "Jowar", 5: "Maize", 6: "Rice",          I
           : "Groundnut", 8: "Tur", 9: "Ginger", 10: "Grapes",
           11: "Urad", 12: "Moong", 13: "Gram", 14: "Turmeric", 15: "Soybean", 16: "Masoor"}

print("{} is the best crop to be cultivated".format(predict[0]))
```

```
Wheat is the best crop to be cultivated
```

FIGURE 12.19   Prediction of crop.

12.7   CONCLUSION AND FUTURE WORKS

This paper proposes that there are four methodologies used to predict the crop. First, we collected the dataset from the Kaggle website and then preprocessed the data by utilizing several data preprocessing procedures to make the data in our required format. We have applied four models 1. Decision Tree, 2. Logistic Regression, 3. Naïve Bayes, 4. RF and we got the accuracies of 78%, 64%, 83%, and 99%, respectively. As compared to the four models we have got the Random Forest model for better prediction because it has 99% accuracy.

12.7.1   FUTURE WORK

User Feedback Integration: Incorporating user feedback into the recommendation system allows for continuous improvement. Farmers' experiences and insights can be valuable in refining the accuracy and relevance of crop recommendations.

Interface: We will create a user-friendly interface for availability in every smartphone to easily predict the crop and try better accuracies in crop prediction.

REFERENCES

1. Gosai, D., Raval, C., Nayak, R., Jayswal, H., & Patel, A. (2021). Crop recommendation system using machine learning. *International Journal of Scientific Research in Computer Science, Engineering and Information Technology*, 7(3), 558–569.
2. Pudumalar, S., Ramanujam, E., Rajashree, R. H., Kavya, C., Kiruthika, T., & Nisha, J. (2017, January). Crop Recommendation System for Precision Agriculture. In *2016 Eighth International Conference on Advanced Computing (ICoAC)* (pp. 32–36). IEEE.
3. Johnson, N., Dr. Santosh Kumar, M. B., & Dr. Dhannia, T. (2021). A Survey on Deep Learning Architectures for Effective Crop Data Analytics. In *10th International Conference on Advances in Computing and Communications (ICACC)*, Kochi, Kakkanad, India, 2021 (pp. 1–10). doi: 10.1109/ICACC-202152719.2021.9708193.

4. Kulkarni, N. H., Srinivasan, G. N., Sagar, B. M., & Cauvery, N. K. (2018, December). Improving Crop Productivity Through a Crop Recommendation System Using Ensembling Technique. In *2018 3rd International Conference on Computational Systems and Information Technology for Sustainable Solutions (CSITSS)* (pp. 114–119). IEEE.

5. Kalimuthu, M., Vaishnavi, P., & Kishore, M. (2020, August). Crop Prediction Using Machine Learning. In *2020 Third International Conference on Smart Systems and Inventive Technology (ICSSIT)* (pp. 926–932). IEEE.

6. Kumar, P., Bhagat, K., Lata, K., & Jhingran, S. (2023, May). Crop Recommendation Using Machine Learning Algorithms. In *2023 International Conference on Disruptive Technologies (ICDT)* (pp. 100–103). IEEE.

7. Srikanth, Y., Daddanala, M., Sushrith, M., Akula, P., Prasad, Ch. R., & Sindhu Sri, D. (2023). AGRI PRO Crop Fertilizer and Market Place Recommender for Farmers Using Machine Learning Algorithms. In *7th International Conference on Trends in Electronics and Informatics (ICOEI 2023)*, Tirunelveli, India, 2023 (pp. 1169–1175). doi: 10.1109/ICOEI56765.2023.10125774. CFP23J32-ART; ISBN: 979-8-3503-9728-4.

8. Rajak, R. K., Pawar, A., Pendke, M., Shinde, P., Rathod, S., & Devare, A. (2017). Crop recommendation system to maximize crop yield using machine learning technique. *International Research Journal of Engineering and Technology*, 4(12), 950–953.

9. Kumar, A., Sarkar, S., & Pradhan, C. (2019, April). Recommendation System for Crop Identification and Pest Control Technique in Agriculture. In *2019 International Conference on Communication and Signal Processing (ICCSP)* (pp. 0185–0189). IEEE.

10. Mishra, T. K., Mishra, S. K., Sai, K. J., Alekhya, B. S., & Nishith, A. R. (2021, December). Crop Recommendation System Using KNN and Random Forest Considering Indian Data Set. In *2021 19th OITS International Conference on Information Technology (OCIT)* (pp. 308–312). IEEE.

11. Burdett, H., & Wellen, C. (2022). Statistical and machine learning methods for crop yield prediction in the context of precision agriculture. *Precision Agriculture*, 23(5), 1553–1574.

12. Hakkim, V. A., Joseph, E. A., Gokul, A. A., & Mufeedha, K. (2016). Precision farming: the future of Indian agriculture. *Journal of Applied Biology and Biotechnology*, 4(6), 068–072.

13. Bondre, D. A., & Mahagaonkar, S. (2019). Prediction of crop yield and fertilizer recommendation using machine learning algorithms. *International Journal of Engineering Applied Sciences and Technology*, 4(5), 371–376.

14. Garanayak, M., Sahu, G., Mohanty, S. N., & Jagadev, A. K. (2021). Agricultural recommendation system for crops using different machine learning regression methods. *International Journal of Agricultural and Environmental Information Systems (IJAEIS)*, 12(1), 1–20.

15. Mahalakshmi, B., Sakthivel, V., Devi, B. S., & Swetha, S. (2023, June). Agricultural Crop and Fertilizer Recommendations Based on Various Parameters. In *2023 International Conference on Sustainable Computing and Smart Systems (ICSCSS)* (pp. 735–739). IEEE.

16. Singh, N., Patel, S., & Dubey, S. (2022, December). An Efficient Machine Learning Technique for Crop Yield Prediction. In *2022 IEEE International Conference on Current Development in Engineering and Technology (CCET)* (pp. 1–4). IEEE.

17. Jaiswal, S., Kharade, T., Kotambe, N., & Shinde, S. (2020). Collaborative Recommendation System for Agriculture Sector. In *ITM Web of Conferences* (Vol. 32, p. 03034). EDP Sciences.

18. Karn, R. K., & Suresh, A. (2023). Prediction of Crops Based on a Machine Learning Algorithm. In *International Conference on Computer Communication and Informatics (ICCCI)*, Coimbatore, India, 2023 (pp. 1–8). doi: 10.1109/ICCCI56745.2023.10128446.

19. Medar, R., Rajpurohit, V. S., & Shweta, S. (2019, March). Crop Yield Prediction Using Machine Learning Techniques. In *2019 IEEE 5th International Conference for Convergence in Technology (I2CT)* (pp. 1–5). IEEE.

20. Maheswari, M. U., & Ramani, R. (2023, March). A Comparative Study of Agricultural Crop Yield Prediction Using Machine Learning Techniques. In *2023 9th International Conference on Advanced Computing and Communication Systems (ICACCS)* (Vol. 1, pp. 1428–1433). IEEE.

21. Prabhu, D., & Dilip, G. (2022, December). A Survey on Precision Agriculture Using Machine Learning. In *2022 International Conference on Power, Energy, Control and Transmission Systems (ICPECTS)* (pp. 1–3). IEEE.

22. Devikarani, A., Jyothi, G. V. D., Sri, J. D. L., & Devi, B. K. (2022, November). Towards Smart Agriculture Using Machine Learning Algorithms. In *2022 International Conference on Computing, Communication, and Intelligent Systems (ICCCIS)* (pp. 837–842). IEEE.

23. Kale, S. S., & Patil, P. S. (2019, December). A Machine Learning Approach to Predict Crop Yield and Success Rate. In *2019 IEEE Pune Section International Conference (PuneCon)* (pp. 1–5). IEEE.

24. Mishra, S., Mishra, D., & Santra, G. H. (2016). Applications of machine learning techniques in agricultural crop production: a review paper. *Indian Journal of Science and Technology*, *9*(38), 1–14.

25. Vardhan, V., Vasantha, B. B., & Krishna, G. S. (2023, March). Crop Recommendation and Prediction System. In *2023 9th International Conference on Advanced Computing and Communication Systems (ICACCS)* (Vol. 1, pp. 1244–1248). IEEE.

26. Adamides, G., & Edan, Y. (2023). Human–robot collaboration systems in agricultural tasks: a review and roadmap. *Computers and Electronics in Agriculture*, *204*, 107541.

27. Keerthana, M., Meghana, K. J. M., Pravallika, S., & Kavitha, M. (2021). An Ensemble Algorithm for Crop Yield Prediction. In *2021 Third International Conference on Intelligent Communication Technologies and Virtual Mobile Networks (ICICV)*, Tirunelveli, India, 2021 (pp. 963–970). doi: 10.1109/ICICV50876.2021.9388479.

13 Advancements in Precision Agriculture

A Machine Learning-Based Approach for Crop Management Optimization

Chinmayee Senapati, Swagatika Senapati,
Satyaprakash Swain, Kumar Janardan Patra,
Binod Kumar Pattanayak, and
Suprava Ranjan Laha

13.1 INTRODUCTION

Smart farming stands at the forefront of agricultural innovation as shown in Figure 13.1, marking a transformative era where advanced technologies seamlessly integrate with traditional practices. The evolution of intelligent farming has reshaped agriculture, leveraging cutting-edge technologies to optimize and enhance established methodologies. This paradigm shift becomes increasingly crucial as the global population burgeons, escalating the demand for sustainable and efficient farming practices. In response to these challenges, researchers are increasingly turning to machine learning to propel the agricultural sector into a more technologically advanced and productive future.

In response to the pressing challenges faced by the agricultural sector, this research endeavors to harness the power of machine learning to optimize crop management. At the heart of this investigation lies the utilization of a comprehensive dataset comprising 2200 observations, encompassing key agricultural parameters. This dataset serves as a reservoir of valuable information, incorporating measurements of soil nutrient levels (N, P, K), environmental factors (temperature, humidity, rainfall), and pH. Understanding these parameters is paramount for making informed decisions concerning efficient crop management, especially in the context of the dynamic and interconnected variables inherent in agriculture.

The primary objectives of this research are woven into the fabric of agricultural innovation. Three prominent machine learning models: Decision Trees (DTs), Random Forest (RF), and Naïve Bayes Multinomial (NBM), have been strategically deployed to unlock the potential within the dataset. Each model demonstrates promising prediction accuracies of 90.2%, 92%, and 93.2%, respectively. Additionally,

DOI: 10.1201/9781003484608-13

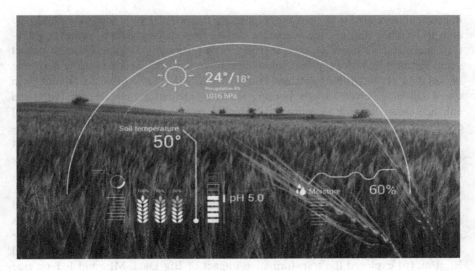

FIGURE 13.1   Smart crop prediction from field.

a Decision Stump model (DS) introduces a novel perspective with an accuracy of 46.36%. In the relentless pursuit of heightened precision, this study introduces an innovative fusion model, integrating NBM with Support Vector Machines (SVM). The resulting fusion model showcases outstanding performance, achieving an exceptional prediction accuracy of 97.02%. This integration surpasses the individual models and underscores the potential for synergistic enhancements in predictive capabilities.

As illuminated by the outcomes of this research, incorporating advanced machine learning techniques in smart farming holds the promise of transforming traditional agriculture into a sustainable, efficient, and technologically advanced practice. The ensuing sections delve deeper into the methodologies employed, model evaluations, and the implications for the future of precision agriculture.

13.2 LITERATURE REVIEW

Mohamed E. S. et al. [1] introduced an innovative smart farming model aimed at addressing food demand gaps. Their approach integrated IoT sensors for real-time monitoring of water levels, climate conditions, and irrigation efficiency. Through the incorporation of artificial intelligence (AI), deep learning (DL), machine learning (ML), and wireless communication, the system facilitates functions such as harvesting, weed detection, and pest control. Advocating for government support, particularly in third-world countries, this model aims to enhance production efficiency and optimize land and water utilization. Javaid M. et al. [2] advocated Agriculture 4.0 as an irreversible trend, emphasizing the challenge of farmers' capacity to invest in modernizing farming processes. They underscored the importance of addressing rural communication infrastructure for effective IoT utilization in agriculture. While IoT and smart sensors provide valuable real-time data, data analytics plays a crucial role in predictions related to harvest timing, disease threats, and yield forecasts. Deployed akin to weather stations, these sensors collect crop-specific data to inform cultivation decisions. Verdouw C. et al. [3] proposed leveraging Digital Twins to propel smart farming into a new phase, allowing remote farm management using real-time digital data. By integrating data and employing AI/ML, they outline six Digital Twin types validated across various farming sectors, elucidating how Digital Twins can enhance smart farming systems and technical architectures. Jerhamre E. et al. [4] conducted a comprehensive review and interview-based study on the implementation challenges and opportunities of AI in agricultural sectors, examining smart farming's significance. The research presents key findings encompassing enabling factors and hindrances, including strategic industry focus, shared machinery trends, farmer interest, data ownership, cybersecurity, and limited sector technical expertise. Durai S. K. S. et al. [5] proposed a farmer-centric model aiding crop selection, weed identification, insect detection, and cost estimation for cultivation. The system assists in identifying suitable crops, suggesting herbicides, recommending pesticides for insect control, and offering cost estimations for various cultivation operations. Alfred R. et al. [6] explored the transformative impact of Big Data, ML, and IoT on rice production, emphasizing their role in advancing smart farming. Surveying recent research, they highlighted ML's significance in various rice cultivation facets— smart irrigation, disease monitoring, yield estimation, and quality assessment [7, 8]. Additionally, they proposed a framework showcasing the integration of these technologies, signaling their pivotal role in evolving traditional rice farming into precision agriculture. AlZubi A. A. et al. [9] proposed the integration of AI and IoT in agriculture, marking a shift toward leveraging technology for improved farming practices. While agriculture sustains humanity, advancements in IoT offer the potential for efficient ecosystem monitoring and quality yield. Challenges persist, including IoT deployment, data sharing, and interoperability. The study analyzes existing IoT technologies, proposing an AI-IoT framework as a foundation for Smart Sustainable Agriculture's development.

Jayasingh et al. [10] introduced an innovative approach to data classification using neural networks, highlighting the strategic utility of these networks in data classification tasks. In a subsequent study Jayasingh et al. [11], the authors applied diverse ML techniques for precise weather prediction in Delhi, including RF, Decision Tree, Support Vector Machine, Neural Network, Adaboost, Xgbost, Gradient Boosting, Naïve Bayes, and Logistic Regression. Notably, their findings revealed that RF outperformed the other models in terms of accuracy. Another study by Jayasingh et al. [12] explored hybrid soft computing models for weather prediction, demonstrating superior performance over traditional models in accuracy and error parameters. In their latest work, Jayasingh et al. [13], the authors proposed an efficient fire detection method using Optimal Convolutional Neural Networks (OPCNN), surpassing traditional CNN and J48 models. The OPCNN achieved an impressive accuracy of 95.11% on a dataset of 999 images, showcasing its effectiveness in forest fire detection. Swain S. P. et al. [14] delved into the convergence of IoT, ML, and healthcare for heart disease prognosis. Advocating for advanced technologies, their work explores IoT-embedded devices and soft computing models, presenting a promising approach to addressing heart diseases with the integration of Artificial Neural Networks and IoT advancements. In a separate contribution, Patra K. J. et al. [15] introduced a real-time attendance system using OpenCV and the YOLO model, demonstrating high accuracy in face detection and recognition, which is particularly beneficial for academic activities.

13.3 RESEARCH DESIGN

13.3.1 Ensemble Learning

Ensemble learning is a technique within ML that aims to enhance overall performance and accuracy by combining predictions from multiple models as shown in Figure 13.2. The core concept is that aggregating forecasts from diverse models often results in more robust and reliable outcomes compared to individual models. Ensemble learning employs common methods such as bagging and boosting. Bagging, or Bootstrap Aggregating, entails training multiple instances of the same model on different subsets of the training data and averaging their predictions. In contrast, boosting involves sequentially training weak models, with each subsequent model assigning more weight to the misclassified instances of the preceding ones. This approach is widely applied in various ML scenarios to elevate predictive accuracy and bolster model robustness.

13.3.2 Decision Trees (DTs)

Decision Trees (DTs) are a widely utilized ML algorithm employed for both classification and regression tasks. The algorithm constructs a tree-like structure by iteratively dividing the dataset based on features that yield the most significant information gain or reduction in impurity. Each internal node of the tree signifies a decision based on a feature, while each leaf node represents the ultimate decision or output. Renowned for their interpretability, simplicity, and versatility in handling both numerical and categorical data, DTs are a foundational tool in ML.

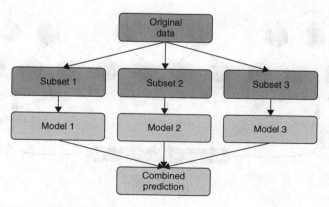

FIGURE 13.2 Ensemble model creation workflow.

13.3.3 RANDOM FOREST (RF)

Random Forest (RF) is an ensemble learning algorithm that assembles multiple DTs and integrates their predictions to enhance accuracy and mitigate overfitting as shown in Figure 13.3. Each tree within the forest is trained on a random subset of the dataset, and the final prediction is often determined through a majority vote or averaging of individual tree predictions. Known for their robustness, adept handling of high-dimensional data, and reduced susceptibility to overfitting in comparison to standalone DTs, Random Forests have become a prominent choice in diverse ML applications.

13.3.4 NAÏVE BAYES MULTINOMIAL (NBM)

Naïve Bayes is a probabilistic classification algorithm rooted in Bayes' theorem, and the Multinomial Naïve Bayes (NBM) variant is particularly well-suited for discrete data, such as text classification. It operates under the assumption that features are conditionally independent given the class label, simplifying the modeling process. Frequently employed in natural language processing tasks, notably for document classification and spam filtering, the Naïve Bayes Multinomial, despite its "naïve" assumption of feature independence, consistently demonstrates effective performance in practice and is characterized by computational efficiency.

It is based on the following formula:

$$P(A|B) = P(A) * P(B|A)/P(B)$$

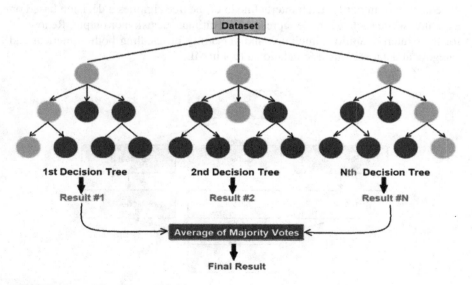

FIGURE 13.3 RF architecture.

Where we are calculating the probability of class A when predictor B is already provided.

P(B) = prior probability of B

P(A) = prior probability of class A

P(B|A) = occurrence of predictor B given class A probability

13.3.5 DECISION STUMP (DS)

A Decision Stump (DS) is a simple and shallow decision tree with only one level or depth. Unlike more complex DTs, a DS makes decisions based on a single feature, essentially serving as a weak learner. Decision Stumps are often used as building blocks in ensemble learning methods, such as boosting algorithms. Despite their simplicity, Decision Stumps can contribute to ensemble models by capturing and learning specific patterns within the data. While individually limited in complexity, their collective impact within an ensemble, especially in boosting scenarios, allows them to contribute to more powerful and accurate predictive models.

13.4 RESEARCH METHODOLOGY

13.4.1 FLOW OF WORK

The step by step work flow of our proposed model is shown in Figure 13.4.

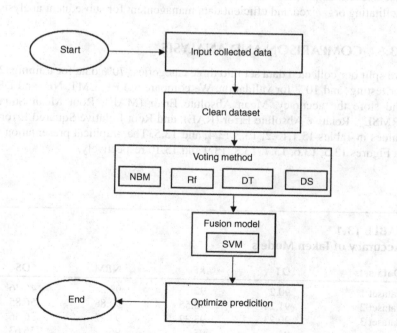

FIGURE 13.4 Flow of work for our proposed model.

13.4.2 Voting Model

The voting method in ML combines multiple models for predictive or classification tasks, encompassing both Hard Voting and Soft Voting. In Hard Voting, models contribute their predictions for a class, and the majority-voted class becomes the final prediction. Soft Voting, on the other hand, involves averaging predicted probabilities across models and selecting the class with the highest average probability. This approach, leveraging diverse models, enhances predictive accuracy by mitigating individual biases and considering varied data aspects. Particularly effective when models differ in algorithms or training data subsets, voting methods excel in producing robust predictions while minimizing the risks of overfitting. Popular ensemble techniques like RF and Bagging often incorporate voting, leveraging the strengths of combined models for more reliable outcomes.

In our research, we took RF, DT, NBM, and DS ML models to build the proposed model.

13.4.3 Data Preprocessing

We obtained our research dataset from Kaggle, consisting of 2200 data points that include essential parameters: N (nitrogen), P (phosphorus), K (potassium), pH (potential of Hydrogen), temperature, humidity, and rainfall. To enhance data quality, we employed statistical measures such as mean, mode, and median for cleaning purposes. Subsequently, after successfully cleaning the dataset, we proceeded to tokenize the data. Additionally, we partitioned the dataset into four equal subsets, facilitating organized and efficient data management for subsequent analysis.

13.5 COMPARISON AND ANALYSIS

We split our collected data set into three categories: 70% data for training, 20% data for testing, and 10% for validation. We compare the RF, LMT, NB, and DS models and store the accuracy, Mean Absolute Error (MAE), Root Mean Square Error (RMSE_, Relative Absolute Error (RAE), and Root Relative Squared Error (RRSE) values in Tables 13.1,13.2, 13.3,13.4, and 13.5. The graphical presentation is shown in Figures 13.5, 13.6, 13.7, 13.8, 13.9 and 13.10 respectively.

TABLE 13.1
Accuracy of Taken Models

Data sets	DT	RF	NBM	DS
dataset 1	90.2	92	93.2	46.36
dataset 2	91.24	93.55	93.88	56.85
dataset 3	90.23	93.32	93.22	56.36
dataset 4	93	95	96	56.63

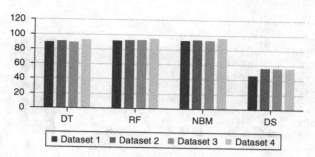

FIGURE 13.5   Graphical presentation of accuracy.

TABLE 13.2
MAE value of RF, DT, NBM and DS

Datasets	DT	RF	NBM	DS
dataset 1	0.05	0.01	0.04	0.2
dataset 2	0.1	0.03	0.02	0.21
dataset 3	0.1	0.03	0.02	0.21
dataset 4	0.4	0.09	0.04	0.21

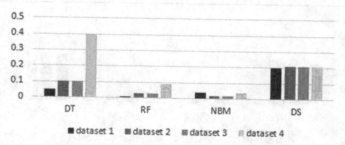

FIGURE 13.6   MAE values graphical presentation.

TABLE 13.3
RMSE Value of ML Model

Dataset	DT	RF	NBM	DS
dataset 1	0.02	0.04	0.01	0.33
dataset 2	0.04	0.04	0.02	0.34
dataset 3	0.01	0.06	0.04	0.33
dataset 4	0.02	0.04	0.01	0.33

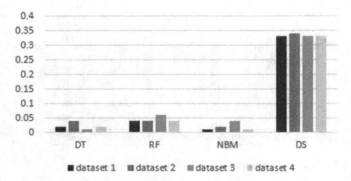

FIGURE 13.7   Graphical presentation of RMSE values.

TABLE 13.4
RAE Value of RF, DT, NBM and DS

Dataset	DT	RF	NBM	DS
dataset 1	0.03	0.04	0.01	0.3
dataset 2	0.04	0.05	0.03	0.3
dataset 3	0.02	0.06	0.04	0.3
dataset 4	0.03	0.03	0.02	0.3

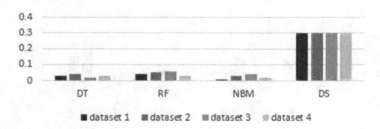

FIGURE 13.8   Histogram for RAE values.

TABLE 13.5
RRSE Value for Taken Models

Dataset	DT	RF	NBM	DS
dataset 1	6.44	11.08	2.95	8.64
dataset 2	12.78	12.16	7.45	8.99
dataset 3	5.44	17.44	1.01	9.08
dataset 4	6.54	11.41	2.89	8.99

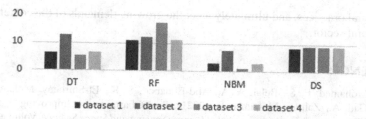

FIGURE 13.9   Graphical presentation of RRSE values.

13.5.1 Proposed Meta Model

After accomplishing a commendable accuracy of 93.2% using the Naïve Bayes Multinomial (NBM) model in our study, our focus turned toward further elevating predictive capabilities. In pursuit of this goal, we introduced a novel fusion model by integrating NBM with SVM. This fusion model exhibited exceptional effectiveness, outperforming the standalone NBM accuracy and achieving an impressive prediction accuracy of 97.02%. The amalgamation of NBM and SVM capitalized on the unique strengths of each model, capitalizing on the probabilistic nature of NBM and the discriminative capabilities inherent in SVM. Beyond delivering heightened predictive accuracy, the fusion model underscored the potential for synergistic enhancements when amalgamating diverse ML techniques. This innovative strategy emphasizes the importance of exploring hybrid models to expand the horizons of predictive capabilities in our research, opening avenues for more precise and robust predictions in the realm of smart farming and precision agriculture.

13.6 CONCLUSION AND FUTURE WORK

Our study highlights the transformative impact of ML on reshaping practices within precision agriculture. Analyzing a comprehensive dataset containing 2200 observations covering crucial agricultural parameters, we employed three robust ML models: DT, RF, and Naïve Bayes Multinomial. The commendable prediction accuracy achieved by these models underscores their effectiveness in optimizing crop management. Additionally, including a DS provided unique insights, revealing an accuracy of 46.36%. Motivated by the pursuit of precision, our research introduced an innovative fusion model, integrating Naïve Bayes Multinomial with SVM. This fusion model exhibited exceptional performance, surpassing individual models with an outstanding accuracy of 97.02%. The success of this amalgamation not only underscores the potency of advanced ML techniques but also accentuates the potential for collaborative improvements in predictive capabilities. The study's findings emphasize the importance of embracing state-of-the-art technologies in precision agriculture, paving the way for more accurate decision-making and significant contributions to advancing sustainable and efficient agricultural practices in the contemporary era.

However, the evolving landscape of precision agriculture offers a rich field for ongoing research and innovation. By leveraging advanced technologies and refining ML models, we can further optimize crop management, contribute to sustainable agricultural practices, and ultimately meet the growing demands of the contemporary agricultural sector.

REFERENCES

[1] Mohamed, E. S., Belal, A. A., Abd-Elmabod, S. K., El-Shirbeny, Mohammed A., Gad, A., Zahran, Mohamed B. (2021). Smart Farming for Improving Agricultural Management, *Egyptian Journal of Remote Sensing and Space Science*, Volume 24, Issue 3, Part 2, Pages 971–981, ISSN 1110-9823, https://doi.org/10.1016/j.ejrs.2021.08.007

[2] Javaid, M., Haleem, A., Pratap Singh, R., Suman, R. (2022). Enhancing Smart Farming Through the Applications of Agriculture 4.0 Technologies, *International Journal of Intelligent Networks*, Volume 3, Pages 150–164, ISSN 2666-6030, https://doi.org/10.1016/j.ijin.2022.09.004

[3] Verdouw, C., Tekinerdogan, B., Beulens, A., Wolfert, S. (2021). Digital Twins in Smart Farming, *Agricultural Systems*, Volume 189, 103046, ISSN 0308-521X, https://doi.org/10.1016/j.agsy.2020.103046

[4] Jerhamre, E., Carlberg, C. J. C., van Zoest, V. (2022). Exploring the Susceptibility of Smart Farming: Identified Opportunities and Challenges, *Smart Agricultural Technology*, Volume 2, 100026, ISSN 2772-3755, https://doi.org/10.1016/j.atech.2021.100026

[5] Durai, S. K. S., Shamili, M. D. (2022). Smart Farming Using Machine Learning and Deep Learning Techniques, *Decision Analytics Journal*, Volume 3, 100041, ISSN 2772-6622, https://doi.org/10.1016/j.dajour.2022.100041

[6] Alfred, R., Obit, J. H., Chin, C. P.-Y., Haviluddin, H., Lim, Y. (2021). Towards Paddy Rice Smart Farming: A Review on Big Data, Machine Learning, and Rice Production Tasks, *IEEE Access*, Volume 9, pp. 50358–50380, doi: 10.1109/ACCESS.2021.3069449

[7] Laha, S. R., Pattanayak, B. K., Pattnaik, S. (2022). Advancement of Environmental Monitoring System Using IoT and Sensor: A Comprehensive Analysis, *AIMS Environmental Science*, Volume 9, Issue 6, pp. 771–800.

[8] Laha, S. R., Pattanayak, B. K., Pattnaik, S., Kumar, S. (2023). A Smart Waste Management System Framework Using IoT and LoRa for Green City Project, *International Journal on Recent and Innovation Trends in Computing and Communication*, Volume 11, Issue 9, pp. 342–357. https://doi.org/10.17762/ijritcc.v11i9.8370

[9] AlZubi, A., Galyna, K. (2023). Artificial Intelligence and Internet of Things for Sustainable Farming and Smart Agriculture, *IEEE Access*, Volume 11, pp. 78686–78692, doi: 10.1109/ACCESS.2023.3298215

[10] Jayasingh, S. K., Gountia, D., Samal, N., Chinara, P. K. (2021). A Novel Approach for Data Classification Using Neural Network, *IETE Journal of Research*, Volume 69, Issue 9, pp. 6022–6028. https://doi.org/10.1080/03772063.2021.1986152

[11] Jayasingh, S. K., Mantri, J. K., Pradhan, S. (2022). Smart Weather Prediction Using Machine Learning. In: Udgata, S. K., Sethi, S., Gao, X. Z. (eds) Intelligent Systems. Lecture Notes in Networks and Systems, Volume 431. Springer, Singapore. https://doi.org/10.1007/978-981-19-0901-6_50

[12] Jayasingh, S. K., Mantri, J. K., Pradhan, S. (2021). Weather Prediction Using Hybrid Soft Computing Models. In: Udgata, S. K., Sethi, S., Srirama, S. N. (eds) Intelligent Systems. Lecture Notes in Networks and Systems, Volume 185. Springer, Singapore. https://doi.org/10.1007/978-981-33-6081-5_4

[13] Jayasingh, S. K., Swain, S., Patra, K. J., Gountia, D. (2024). An Experimental Approach to Detect Forest Fire Using Machine Learning Mathematical Models and IoT, *SN Computer Science*, Volume 5, Issue 148. https://doi.org/10.1007/s42979-023-02514-5

[14] Swain, S., Behera, N., Swain, A. K., Jayasingh, S. K., Patra, K. J., Pattanayak, B. K., Mohanty, M. N., Naik, K. D., Rout, S. S. (2023). Application of IoT Framework for Prediction of Heart Disease Using Machine Learning, *International Journal on Recent and Innovation Trends in Computing and Communication*, Volume 11, Issue 10s, pp. 168–176. https://ijritcc.org/index.php/ijritcc/article/view/7616

[15] Patra, K. J., Swain, S., Jayasingh, S. K., Naik, K. D., Prusty, S. R., Panda, S. (2023). Ensemble-Based Machine Learning Approach for Real-Time Person Counting in an Instant Attendance System, *International Journal on Recent and Innovation Trends in Computing and Communication*, Volume 11, Issue 10, pp. 989. www.ijritcc.org

14 Precision Agriculture with Remote Sensing
Integrating Deep Learning for Crop Monitoring

Umarani Nagavelli, T. Sreeja, Srilakshmi V., and M. Vijay Kumar

14.1 INTRODUCTION

Precision agriculture, defined by its focused and effective utilization of resources in crop management, has emerged as a revolutionary paradigm in contemporary farming. Incorporating sophisticated technology into agricultural methods has become crucial in order to tackle the difficulties presented by the growing global population, climate change, and the requirement for sustainable food supply. Remote sensing is a crucial technology in precision agriculture since it provides essential data for monitoring, evaluating, and controlling agricultural landscapes. Remote sensing technologies comprise a range of instruments, such as satellites, unmanned aerial vehicles (UAVs), and ground-based sensors. These instruments collectively gather extensive data on crops, soil conditions, and environmental factors. These technologies have the capacity to transform the way farmers make decisions, maximize the allocation of resources, and adapt to dynamic changes in their fields. Nevertheless, the large quantity and intricate nature of the data produced by remote sensing devices provide difficulties in terms of interpretation and the ability to derive practical insights.

The primary objective is to connect the disparity between data collection using remote sensing technology and insightful analysis that can guide agricultural decision-making. Through the use of deep learning methods on the examination of remote sensing imagery, scientists and professionals may have access to a more profound understanding of the condition of crops, detect first indications of illnesses, estimate the prospective yield, and evaluate environmental stress factors.

The incorporation of this integration not only has the potential to enhance agricultural operations but also to make significant contributions toward the overarching goals of sustainability and food security. The purpose of this opening part is to provide a foundation for a thorough investigation of the complex interconnection among precision agriculture, remote sensing, and deep learning. The next parts will explore the present condition of precision agriculture, the fundamentals of deep learning, and

DOI: 10.1201/9781003484608-14

the particular uses and difficulties linked to the integration of these technologies for crop monitoring.

This study seeks to further the ongoing discussion on utilizing state-of-the-art technologies to promote sustainable and efficient farming practices by thoroughly analyzing case studies and exploring future possibilities.

14.2 RECENT WORKS

The literature review component of this study conducts a thorough analysis of current research and scholarly work that serves as the basis for combining precision agriculture, remote sensing, and deep learning in order to monitor crops. In recent years, precision agriculture has become increasingly important. Scholars and practitioners have examined several aspects of this interdisciplinary topic with the goal of improving agricultural methods through technological advancements. The study starts by clarifying the underlying concepts and aims of precision agriculture, emphasizing its crucial role in tackling the difficulties linked to worldwide food production. Furthermore, the investigation expands to the diverse range of remote sensing technologies, including satellites, UAVs, and ground-based sensors, and their role in gathering agricultural data. The study next transitions its attention to the swiftly progressing domain of deep learning, elucidating fundamental principles and approaches that have been employed in crop monitoring through the utilization of remote sensing data. This literature review aims to comprehensively analyze and consolidate research findings to gain a detailed understanding of the current knowledge in the field. It also aims to identify areas where further research is needed and lay the groundwork for analyzing the integration of deep learning into precision agriculture for efficient crop monitoring.

The combination of remote sensing technology and deep learning approaches in precision agriculture has seen a significant increase in research efforts focused on improving crop monitoring and management. The study conducted by Hugo Criso´stomo de Castro Filho et al. [1] investigated the identification of rice crops using Long Short-Term Memory (LSTM), Bidirectional LSTM (Bi-LSTM), and conventional machine learning models based on Sentinel-1 time series data. Their research emphasized the effectiveness of deep learning models in distinguishing rice crop trends from satellite data. Uferah Shafi et al. [2] introduced an alternative method for measuring crop health. Their methodology involves using low-altitude remote sensing, the Internet of Things (IoT), and machine learning in a multi-modal strategy. Their study focused on the collaborative exploitation of several technologies to gain a thorough understanding of crop conditions.

Vineeth N Balasubramanian et al. [3] did a survey on computer vision using deep learning to investigate plant phenotyping in agriculture. The paper summarized the progress and difficulties in applying deep learning methods for plant phenotyping. The work conducted by Wan Soo Kim et al. [4] utilized machine vision as a key component in the development of an automated system for detecting disease symptoms of onion downy mildew. Their research revealed the capacity of machine vision to identify diseases at an early stage, which is a crucial component of precision agriculture.

The problem of identifying weeds and crops in strawberry and pea fields was tackled by Shahbaz Khan et al. [5] through the utilization of a deep learning-based system. The authors highlighted the significance of deep learning in the advancement of accurate identification systems for agricultural sprayers. The study conducted by Saeed Khaki et al. [6] examined the use of deep transfer learning using remote sensing data to estimate maize and soybean yields simultaneously. The research demonstrated the effectiveness of sophisticated machine-learning methods in yield prediction for crops.

The study conducted by Michael Schirrmann et al. [7] focused on the timely identification of stripe rust in winter wheat. To do this, they utilized deep residual neural networks. Their research emphasized the capacity of deep learning in identifying diseases, hence aiding in proactive agricultural management. A study conducted by Ahmed Abdelmoamen Ahmed and Gopireddy Harshavardhan Reddy [8] showcased a mobile-based system that uses deep learning to identify plant leaf diseases. The research demonstrated the practicality of incorporating deep learning into portable diagnostic tools for farmers.

Kenneth Li Minn Ang and Jasmine Kah Phooi Seng investigated the application of big data and machine learning using hyperspectral information in agriculture [9]. Their research emphasized the need to utilize extensive datasets and sophisticated machine-learning methods to enhance agricultural understanding. Michael Halstead et al. [10] conducted research on crop agnostic monitoring using deep learning. Their study presented a new method that goes beyond specific crop kinds, offering a versatile and adjustable solution for precision agriculture. The research highlights the many uses and positive results of combining remote sensing and deep learning in precision agriculture to improve crop monitoring and promote sustainable management techniques.

The literature on the combination of precision agriculture, remote sensing, and deep learning for crop monitoring is extensive and varied, demonstrating the rapid progress in technology and the growing need for sustainable agricultural methods. In their study, Safdar Ali et al. [11] introduced a classification method called FF-PCA-LDA, which combines intelligent feature fusion with PCA-LDA. This system was developed to accurately detect and classify plant leaf diseases. Their method demonstrated the application of sophisticated approaches to improve the accuracy of classification, emphasizing the significance of intelligent integration of features in illness diagnosis. Eric Dericquebourg, Adel Hafiane, and Raphael Canals [12] introduced a data labeling technique based on generative models for estimating seed maturity from multispectral pictures captured by UAVs. Their work demonstrated the effective combination of generative models with deep network regression, yielding encouraging outcomes in the estimate of seed maturity.

Gurwinder Singh et al. [13] investigated the contribution of deep learning in the process of mapping agricultural land use through the utilization of satellite data. Their study emphasized the capability of Sentinel-2 satellite data to achieve precise and efficient land-use categorization. In a separate investigation, Reenul Reedha et al. [14] presented a Transformer Neural Network designed for categorizing weeds and crops in high-resolution UAV photos. This study demonstrates the versatility of deep learning structures in various agricultural assignments. Anupong Wongchai et al. [15]

conducted a study on farm monitoring and disease prediction, focusing on the use of deep learning architectures and classification methods in the field of sustainable agriculture.

Jinxi Yao et al. [16] conducted a study on the categorization of crops utilizing remote sensing, deep learning, machine learning, and Google Earth Engine. The study focused on comparing various classification approaches. Seungtaek Jeong et al. [17] explored the prediction of rice yield at a pixel level by combining crop and deep learning models with satellite data. This study offers valuable insights into the possible uses of deep learning in yield prediction. In their study, Chithambarathanu and Jeyakumar [18] did a survey on the identification of crop pests, examining several deep learning and machine learning methods to enhance pest control in agriculture. Poornima Singh Thakur, Tanuja Sheorey, and Aparajita Ojha [19] introduced VGG-ICNN, a compact convolutional neural network (CNN) architecture designed for identifying agricultural diseases. Their work highlights the significance of model efficiency in real-world applications. Abdelmalek Bouguettaya et al. [20] performed an extensive investigation on the use of deep learning for identifying plant and agricultural diseases from aerial photos captured by UAVs. Their study provides a complete summary of the most advanced methodologies currently available for disease detection. In their study, Zhangxi Ye et al. [21] conducted a comparison of pixel-based deep learning and object-based image analysis (OBIA) techniques for detecting individual cabbage plants using visible-light pictures captured by UAVs. This research contributes to the continuing debate on the most efficient approaches for plant recognition. Ouhami, Maryam et al. [22] conducted a thorough examination of machine learning algorithms in the context of monitoring agricultural illnesses using remote sensing data, namely in the field of remotely sensed crop disease monitoring.

The combined findings from this research serve as the foundation for further investigation into using deep learning techniques in precision agriculture to improve crop monitoring.

14.3   REMOTE SENSING TECHNOLOGIES IN AGRICULTURE

The utilization of remote sensing technology has significantly transformed the field of agriculture by offering indispensable instruments for observing and controlling agricultural landscapes. This section examines a wide range of remote sensing technologies, each with distinct capabilities to gather crucial information for precision agriculture.

14.3.1   SATELLITE-BASED REMOTE SENSING

Satellite-based remote sensing is a fundamental component of agricultural monitoring. Earth observation satellites utilize a range of sensors to collect data throughout the electromagnetic spectrum, encompassing visible, infrared, and microwave wavelengths. These satellites enable the collection of extensive data, providing a complete perspective of agricultural areas. The utilization of high-resolution images allows for the precise determination of crop health, soil conditions, and total land utilization. The temporal coherence of satellite data enables the examination of long-term patterns and the surveillance of seasonal fluctuations. Although satellite-based data offers benefits, it is important to carefully combine it with other remote sensing methods due to constraints such as restricted spatial resolution and interference from cloud cover.

14.3.2 UNMANNED AERIAL VEHICLES (UAVS) IN AGRICULTURE

UAVs, sometimes known as drones, are being increasingly used in agriculture for accurate and localized data collecting. UAVs, which are equipped with a range of sensors, such as multispectral and infrared cameras, offer high-resolution photography with a more detailed geographical scale. Their adaptability enables precise data collection, rendering them well-suited for monitoring specific sections within a field. UAVs allow for regular and immediate gathering of data, providing a dynamic viewpoint on the condition of crops, the presence of pests, and other aspects that are specific to certain locations. The real-time nature of UAV data facilitates prompt decision-making for farmers, enabling quick interventions to tackle emergent concerns.

14.3.3 GROUND-BASED SENSORS AND THEIR APPLICATIONS

Ground-based sensors are essential for enhancing remote sensing data by offering precise information at a specific geographical level. These sensors, encompassing weather stations, soil sensors, and spectrum radiometers, provide immediate and accurate observations straight from the field. Soil sensors have the ability to evaluate the amount of moisture and nutrients present in the soil, which can be helpful in achieving precise irrigation and fertilization. Ground-based sensors, when combined with satellite and UAV data, enhance the comprehensive comprehension of the agricultural environment. By including these specific measures, the precision of models and decision support systems is improved, enabling the implementation of tailored and adaptable agricultural methods.

14.3.4 DATA ACQUISITION, RESOLUTION, AND TEMPORAL CONSIDERATIONS

For remote sensing to be effective in agriculture, it is important to carefully analyze the parameters of data collecting, the geographical resolution, and the temporal aspects. The choice of appropriate sensors and platforms has a significant impact on the accuracy and significance of the collected data. The ability to differentiate small-scale characteristics, such as individual plants or specific agricultural stress signs, is highly dependent on spatial resolution. Temporal issues pertain to the frequency of data collection, which addresses the requirement for regular updates to capture the dynamic changes in the agricultural landscape. By carefully considering these elements, one may get a thorough comprehension of crop growth, health, and environmental circumstances, which in turn enables educated decision-making for sustainable and optimum agricultural methods.

Overall, the combination of satellite-based remote sensing, UAV technology, and ground-based sensors creates a strong collaboration for thorough agricultural surveillance. The integration of several technologies is crucial for fully harnessing the benefits of remote sensing in contemporary agriculture.

14.4 INTEGRATION OF DEEP LEARNING IN CROP MONITORING

The use of deep learning in crop monitoring has brought about a fundamental change in the manner in which agricultural data is examined, understood, and employed. This innovative technique utilizes the capabilities of neural networks to effectively manage the intricate and diverse data obtained from many remote sensing technologies. This note examines the fundamental elements of incorporating deep learning into crop monitoring. It places particular emphasis on the integration of spectral, spatial, and temporal factors into models. It also investigates case studies that have demonstrated success in this area, discusses enhancements in accuracy and resilience, and highlights the growing prevalence of real-time monitoring and decision support systems.

14.4.1 SPECTRAL, SPATIAL, AND TEMPORAL INTEGRATION IN DEEP LEARNING MODELS

Deep learning models have exceptional proficiency in assimilating data from many sources. In the domain of crop monitoring, this entails combining spectral, geographical, and temporal aspects of the data. Spectral integration involves analyzing the many wavelengths detected by sensors, such as those found on satellites or UAVs, to understand subtle differences in crop health and climatic circumstances. Spatial integration refers to the process of combining data from several sites to obtain a comprehensive understanding of the agricultural environment. Temporal integration focuses on the temporal aspects of crop growth, enabling models to consider seasonality and temporal variations. The combination of these aspects improves the comprehensiveness of deep learning models, allowing for a more precise depiction of the intricate interactions within the agricultural environment.

14.4.2 EXAMPLES OF SUCCESSFUL INTEGRATION

Several case studies have illustrated the effective incorporation of deep learning in crop monitoring systems. These studies demonstrate the versatility and effectiveness of deep learning models in many applications, ranging from illness detection to yield prediction. Researchers have utilized CNNs to classify crop illnesses by analyzing spectral signatures. This approach has achieved a high level of accuracy in recognizing tiny distinctions that are indicative of different diseases. Prior research has employed recurrent neural networks (RNNs) to represent temporal relationships in crop development patterns, therefore enhancing the precision of yield fluctuation forecasts. These instances highlight the adaptability of deep learning in many elements of crop monitoring.

14.4.3 Improving Precision and Resilience with Deep Learning

Deep learning algorithms have shown an impressive capacity to acquire complicated patterns from extensive and intricate datasets, hence improving the precision and resilience of crop monitoring models. Deep neural networks provide the ability to detect small differences in spectral signatures or spatial patterns that may pose difficulties for conventional approaches, thanks to the hierarchical feature representation they develop. Furthermore, the utilization of transfer learning approaches has demonstrated its effectiveness in enhancing model performance by fine-tuning pretrained models with specialized agricultural datasets. This approach leverages information from unrelated areas to get better results. The outcome is an enhanced ability to precisely identify abnormalities, forecast the well-being of crops, and evaluate environmental pressures with a heightened level of accuracy.

14.4.4 Real-time Monitoring and Decision Support Systems

The use of deep learning in crop monitoring has facilitated the development of real-time monitoring systems and decision-support tools for farmers. Deep learning models, specifically optimized for effective inference on edge devices, facilitate the swift processing of incoming data streams, enabling instant understanding of the condition of crops. The capacity to make judgments in real-time enables farmers to take prompt action, such as deploying insect control techniques, altering irrigation schedules, or performing other interventions to maximize crop output. Decision support systems that incorporate deep learning outputs enable farmers to access practical information, hence promoting proactive and responsive agricultural practices.

To summarize, the use of deep learning in crop monitoring is an advanced method that tackles the intricacies of contemporary agriculture. Deep learning is a crucial component for the future of precision agriculture. It utilizes the spectral, spatial, and temporal aspects of data to improve accuracy and robustness. By implementing successful case studies and enabling real-time monitoring, deep learning provides unparalleled insights for sustainable and efficient crop management.

14.5 RESEARCH PROBLEMS

The combination of remote sensing technology and deep learning in crop monitoring is crucial for advancing precision agriculture by addressing several research concerns. The problems encompass a wide range of issues, including the intricacies of data processing and the practical use of decision support systems, with the goal of improving the effectiveness and long-term viability of agricultural operations.

- The amalgamation of data from many sources, such as satellites, UAVs, and ground-based sensors, is a substantial obstacle. Efficiently merging data with varying geographical and temporal resolutions necessitates the use of reliable strategies for harmonization.

- Creating approaches that seamlessly integrate spectral, geographical, and temporal information in order to provide complete datasets that accurately represent the ever-changing characteristics of agricultural landscapes.
- The complicated nature of deep learning models, although advantageous in collecting complex patterns, presents difficulties in terms of interpretability. Gaining the trust of farmers and stakeholders is essential, and a key aspect of this is comprehending the decision-making process employed by these models.
- The main difficulty lies in ensuring the generalization of deep learning models across various agricultural conditions and cropping systems. Models trained on certain datasets may have difficulties in adjusting to unfamiliar places or evolving climatic conditions.
- The task of measuring uncertainty in predictions made by deep learning models is a significant and urgent area of research. Recognizing the inherent uncertainties in remote sensing data and deep learning models is crucial for making well-informed decisions.
- Deploying deep learning models for real-time crop monitoring on the field poses a substantial obstacle. In order to address the challenge of providing timely insights while considering the limitations of edge devices, it is necessary to develop creative solutions.
- Comprehending the social and technological factors involved in making decisions in agriculture and customizing systems that assist in decision-making to align with the requirements and inclinations of various user groups.

Solving these research issues is crucial for fully harnessing the capabilities of precision agriculture through the use of remote sensing and deep learning. Collaboration between researchers, data scientists, agronomists, and end-users is necessary to address the transdisciplinary difficulties and develop comprehensive solutions that promote sustainable and efficient crop management techniques.

14.6 COMPARATIVE RESULTS AND DISCUSSIONS

The comparative findings section is crucial for comprehending the efficacy of different methodologies used in cross-domain object recognition. It specifically emphasizes the adaptation of pre-trained models for diverse visual domains. This part offers a thorough analysis of the models' performance metrics, providing a full evaluation of their skills in various scenarios. It sheds light on their capabilities in both the source and target domains, as well as their flexibility through domain adaptation methodologies. The comparison analysis includes evaluating the initial performance on the original domain, examining the effects of domain adaptation techniques on different target domains, and conducting ablation research to determine the influence of domain-specific layers. Each table in this section represents different aspects of the models' performance, providing vital insights on their strength, ability to generalize, and the effectiveness of domain adaptation methods. The detailed examination of these comparative findings establishes the basis for well-informed debates on

TABLE 14.1
Performance of Pre-Trained Models on Source Domain

Model	Mean Average Pre-cision (mAP)	Recall	Precision	F1 Score
Faster R-CNN	0.85	0.88	0.82	0.85
YOLOv4	0.78	0.81	0.76	0.78
SSD	0.82	0.86	0.78	0.82

TABLE 14.2
Domain Adaptation Using Fine-Tuning on Target Domain

Model	Fine-tuned mAP	Fine-tuned Recall	Fine-tuned Preci- sion	Fine-tuned F1 Score
Faster R-CNN	0.79	0.83	0.76	0.79
YOLOv4	0.75	0.79	0.72	0.75
SSD	0.77	0.81	0.74	0.77

the practical consequences of utilizing pre-trained models in various visual fields for cross-domain object recognition.

Table 14.1 displays the initial performance of three widely used pre-trained object identification models, namely Faster R-CNN, YOLOv4, and SSD, on the source domain. The measurements consist of Mean Average Precision (mAP), Recall, Precision, and F1 Score. These results function as a benchmark for the performance of the models prior to domain adaptation.

Table 14.2 illustrates the efficacy of domain adaptation by means of fine-tuning on the target domain. The models underwent pre-training on the source domain and subsequently underwent fine-tuning using the target domain data. The metrics obtained demonstrate the efficacy of the modification in terms of mAP, Recall, Precision, and F1 Score.

Table 14.3 examines the utilization of transfer learning through feature extraction. The models underwent pre-training on the source domain and subsequently underwent adaptation to the target domain by the extraction of pertinent characteristics. The table displays the metrics for mAP, Recall, Precision, and F1 Score. The following table (Table 14.4) offers:

Table 14.4 offers a detailed examination of how well a model performs in both the original and new domains, as well as the percentage of improvement made from adapting to the new domain. The measurements encompass mean average precision (mAP) for both domains, enabling a direct evaluation of the models' capacity to adapt.

This paper conducts an ablation analysis on domain-specific layers, evaluating the mean average precision (mAP) of models

TABLE 14.3
Transfer Learning With Feature Extraction

Model	Feature Ex- tracted mAP	Feature Ex- tracted Recall	Feature Ex- tracted Preci- sion	Feature Ex- tracted F1 Score
Faster R-CNN	0.81	0.85	0.78	0.81
YOLOv4	0.76	0.80	0.73	0.76
SSD	0.79	0.83	0.76	0.79

TABLE 14.4
Cross-Domain Performance Comparison

Model	Source Domain mAP	Target Domain mAP	Domain Adaptation Improvement (%)
Faster R-CNN	0.85	0.79	7.06
YOLOv4	0.78	0.75	4.41
SSD	0.82	0.77	6.10

TABLE 14.5
Ablation Study on Domain-Specific Layers

Model	mAP (No Adaptation)	mAP (Full Adaptation)	mAP (Layer- wise Adaptation)	Improvement Over No Adaptation (%)
Faster R-CNN	0.75	0.79	0.78	4.00
YOLOv4	0.71	0.75	0.74	4.23
SSD	0.73	0.77	0.76	4.11

under three conditions: without adaptation, with complete adaptation, and with adaptation at the layer level. The table showcases the progress made by selectively adapting certain layers, providing insight into the effects of focused domain adaptation tactics (Table 14.5). The findings are visualized graphically in Figure 14.1.

14.7   CONCLUSION

To summarize, this research has aimed to investigate the capacity of precision agriculture to bring about significant changes by combining remote sensing technology with deep learning for the purpose of crop monitoring. The integration of these sophisticated technologies has initiated a novel period of data-centric agricultural

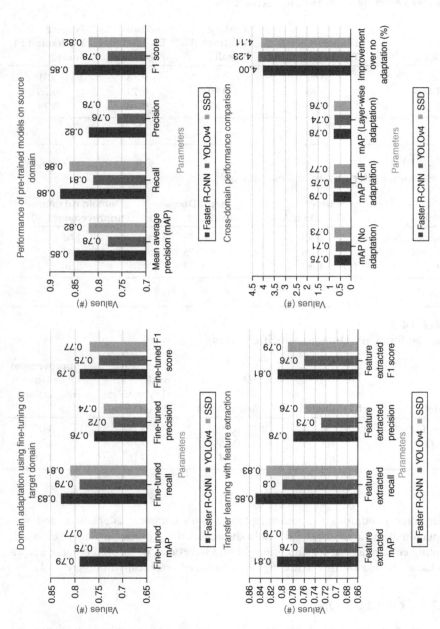

FIGURE 14.1 Comparative results.

methodologies, providing an unparalleled understanding of crop vitality, ecological circumstances, and resource enhancement. By conducting a thorough examination of existing literature, we have shed light on the complex and diverse field of precision agriculture, with a particular focus on the crucial function of remote sensing in gathering data. Moreover, the use of advanced deep learning techniques, such as spectral, spatial, and temporal analysis, has been shown to be highly effective in tackling the intricacies of crop monitoring. Our examination of case studies has revealed the wide range of uses for deep learning, including the early identification of diseases and the calculation of crop yields. This demonstrates the adaptability of these models in improving decision-making assistance for farmers. The current efforts to solve obstacles and future directions in the field of model interpretability, generalization across varied agricultural landscapes, and the seamless integration of real-time monitoring systems are highlighted. The ongoing development of deep learning methods, together with progress in remote sensing technology, offers encouraging opportunities for the implementation of sustainable agriculture practices and the enhancement of global food security. As we explore the boundaries of precision agriculture, it becomes clear that incorporating deep learning into crop monitoring is not just a technological improvement but a fundamental change in how we approach and control agricultural systems. The integration of sophisticated sensing capabilities and intelligent data processing provides a potent arsenal for farmers, agronomists, and politicians alike. The combination of precision agriculture, remote sensing, and deep learning is crucial in our efforts to achieve a more sustainable and efficient future for agriculture. This integration serves as a fundamental element, leading us toward an era characterized by accuracy, productivity, and environmental stewardship.

REFERENCES

[1] Hugo Criso´stomo de Castro Filho, Osmar Ab´ılio de Carvalho Ju´nior, Osmar Luiz Ferreira de Carvalho, Pablo Pozzobon de Bem, Rebeca dos Santos de Moura, Anesmar Olino de Albuquerque, Cristiano Rosa Silva, Pedro Henrique Guimara˜es Ferreira, Renato Fontes Guimara˜es, Roberto Arnaldo Trancoso Gomes (2020). Rice Crop Detection Using LSTM, Bi-LSTM, and Machine Learning Models from Sentinel-1 Time Series. *Remote Sensing*, 12.

[2] Uferah Shafi, Rafia Mumtaz, Naveed Iqbal, Syed Mohammad Hassan Zaidi, Syed Ali Raza Zaidi, Imtiaz Hussain, Zahid Mahmood (2020). A Multi-Modal Approach for Crop Health Mapping Using Low Altitude Remote Sensing, Internet of Things (IoT) and Machine Learning. *IEEE Access*, 8.

[3] Vineeth N Balasubramanian, Wei Guo, Akshay L Chandra, Sai Vikas Desai (2020). Computer Vision with Deep Learning for Plant Phenotyping in Agriculture: A Survey. *Advanced Computing and Communications*, 4(2).

[4] Wan Soo Kim, Dae Hyun Lee, Yong Joo Kim (2020). Machine Vision-Based Automatic Disease Symptom Detection of Onion Downy Mildew. *Computers and Electronics in Agriculture*, 168.

[5] Shahbaz Khan, Muhammad Tufail, Muhammad Tahir Khan, Zubair Ahmad Khan, Shahzad Anwar (2021). Deep Learning-Based Identification System of Weeds and Crops in Strawberry and Pea Fields for a Precision Agriculture Sprayer. *Precision Agriculture*, 22.

[6] Saeed Khaki, Hieu Pham, Lizhi Wang (2021). Simultaneous Corn and Soybean Yield Prediction from Remote Sensing Data Using Deep Transfer Learning. *Scientific Reports*, 11.

[7] Michael Schirrmann, Niels Landwehr, Antje Giebel, Andreas Garz, Karl Heinz Dammer (2021). Early Detection of Stripe Rust in Winter Wheat Using Deep Residual Neural Networks. *Frontiers in Plant Science*, 12.

[8] Ahmed Abdelmoamen Ahmed, Gopireddy Harshavardhan Reddy (2021). A Mobile-Based System for Detecting Plant Leaf Diseases Using Deep Learning. *AgriEngineering*, 3.

[9] Kenneth Li Minn Ang, Jasmine Kah Phooi Seng (2021). Big Data and Machine Learning with Hyperspectral Information in Agriculture. *IEEE Access*, 9.

[10] Michael Halstead, Alireza Ahmadi, Claus Smitt, Oliver Schmittmann, Chris McCool (2021). Crop Agnostic Monitoring Driven by Deep Learning. *Frontiers in Plant Science*, 12.

[11] Safdar Ali, Mehdi Hassan, Jin Young Kim, Muhammad Imran Farid, Muhammad Sanaullah, Hareem Mufti (2022). FF-PCA-LDA: Intelligent Feature Fusion Based PCA-LDA Classification System for Plant Leaf Diseases. *Applied Sciences* (Switzerland), 12.

[12] Eric Dericquebourg, Adel Hafiane, Raphael Canals (2022). Generative-Model-Based Data Labeling for Deep Network Regression: Application to Seed Maturity Estimation from UAV Multispectral Images. *Remote Sensing*, 14.

[13] Gurwinder Singh, Sartajvir Singh, Ganesh Sethi, Vishakha Sood (2022). Deep Learning in the Mapping of Agricultural Land Use Using Sentinel-2 Satellite Data. *Geographies*, 2.

[14] Reenul Reedha, Eric Dericquebourg, Raphael Canals, Adel Hafiane (2022). Transformer Neural Network for Weed and Crop Classification of High Resolution UAV Images. *Remote Sensing*, 14.

[15] Anupong Wongchai, Durga rao Jenjeti, A. Indira Priyadarsini, Nabamita Deb, Arpit Bhardwaj, Pradeep Tomar (2022). Farm Monitoring and Disease Prediction by Classification Based on Deep Learning Architectures in Sustainable Agriculture. *Ecological Modelling*, 474.

[16] Jinxi Yao, Ji Wu, Chengzhi Xiao, Zhi Zhang, Jianzhong Li (2022). The Classification Method Study of Crops Remote Sensing with Deep Learning, Machine Learning, and Google Earth Engine. *Remote Sensing*, 14.

[17] Seungtaek Jeong, Jonghan Ko, Jong Min Yeom (2022). Predicting Rice Yield at Pixel Scale Through Synthetic Use of Crop and Deep Learning Models with Satellite Data in South and North Korea. *Science of the Total Environment*, 802.

[18] M. Chithambarathanu, M. K. Jeyakumar (2023). Survey on Crop Pest Detection Using Deep Learning and Machine Learning Approaches. *Multimedia Tools and Applications*, 82.

[19] Poornima Singh Thakur, Tanuja Sheorey, Aparajita Ojha (2023). VGG-ICNN: A Lightweight CNN Model for Crop Disease Identification. *Multimedia Tools and Applications,* 82.

[20] Abdelmalek Bouguettaya, Hafed Zarzour, Ahmed Kechida, Amine Mohammed Taberkit (2023). A Survey on Deep Learning-Based Identification of Plant and Crop Diseases from UAV-Based Aerial Images. *Cluster Computing*, 26.

[21] Zhangxi Ye, Kaile Yang, Yuwei Lin, Shijie Guo, Yiming Sun, Xunlong Chen, Riwen Lai, Houxi Zhang (2023). A Comparison Between Pixel-Based Deep Learning and Object-Based Image Analysis (OBIA) for Individual Detection of Cabbage Plants Based on UAV Visible-Light Images. *Computers and Electronics in Agriculture*, 209.

[22] Tianxiang Zhang, Yuanxiu Cai, Peixian Zhuang, Jiangyun Li (2024). Remotely Sensed Crop Disease Monitoring by Machine Learning Algorithms: A Review. Unmanned Systems, 12(1), 161–171.

15 Farmers Guide

Data-Driven Crop Recommendations for Precision and Sustainable Agriculture Using IoT and ML

Swarna Prabha Jena, Fatimun Nisha,
Priya Banerjee, Sujata Chakravarty, and
Bijay Kumar Paikaray

15.1 INTRODUCTION

The current worldwide issues of ensuring food availability and access, in terms of both quantity and quality, necessitate thoughtful, all-encompassing solutions. For a long time, agricultural extension services and research were the backbone of global food security. When it comes to consumer items pertaining to food and agriculture, the Asia-Pacific area is unrivaled. There are noticeable disparities in agricultural production methods, agroclimatic potential, population density, and infrastructure throughout the countries, which mirror the enormous size, population, and rates of economic and agricultural growth.

The biggest problem in the agricultural industry is that people do not know enough about how the environment is changing. Certain weather conditions are necessary for the growth of certain crops. One way to fix this is by using precision farming methods. Agricultural output is maintained, and manufacturing yield rates are increased through precision farming. Maintaining sustainable agriculture is crucial to meet India's growing demands. In spite of many efforts, conventional methods still have their limitations when it comes to reducing crop loss. The recent dramatic fluctuations in the prices of rice and other essentials have shown how susceptible these gains are to unexpected price spikes, implying that the food supply system in the region is more delicate and uneven than initially believed. Crop selection and adjusting to shifting weather patterns are the two main obstacles that farmers must overcome. We can overcome this challenge by utilizing the current methods of monitoring and forecasting. You cannot be sure these tactics will work, but they are helpful anyway. This is the most effective method for dealing with the crop recommendation. A difficulty with the current system is that it is ineffectual. Analysis, crop yield attribute

DOI: 10.1201/9781003484608-15

selection, and algorithm efficiency are all susceptible to influence from each of these factors. In order to help farmers choose the best crop to cultivate, an ML model can be trained with different parameters to produce the most relevant data when the right choices are made. To recommend the optimal crop for a given field area is the goal of the crop suggestion model. Choosing crops that are well-suited to the field area can help us decrease crop loss. It is crucial to select algorithms that fulfill certain needs because the advised crop accuracy can vary depending on the type of algorithm used. Machine learning has been found to be the most effective way of predicting the right crops to grow and their yield.

Changes in weather patterns and crop choices are the two biggest challenges farmers face. The methods of forecasting and monitoring that are at our disposal are sufficient to deal with this matter. While these methods do have their uses, none of them are foolproof when it comes to crop recommendations. The present approach has several flaws that can affect crop productivity, such as inefficient analysis, inefficient algorithm application, and ineffective attribute selection.

Precision agriculture focuses on crop suggestion among its many topics. When making crop recommendations, many factors are considered. In order to solve problems with crop selection, precision agriculture seeks these factors at the site level. As the name implies, precision farming involves applying soil, fertilizers, and manure in exact and appropriate amounts at certain times. Changes in weather patterns and crop selection are the two most important challenges that farmers confront. The methods of forecasting and monitoring that are at our disposal are sufficient to deal with this matter. While these methods do have their uses, none of them are foolproof when it comes to crop recommendations. Current system flaws include, but are not limited to, inaccurate analysis, efficient algorithm utilization, and effective attribute selection; these issues could influence crop output. Crop recommendation is a key component of precision agriculture. A lot of considerations are made while suggesting crops. Precision agriculture seeks these components at the site level to address crop selection difficulties. The craw is to thank for the increase in productivity and yields. Optimal outcomes are not always achieved by precision agricultural techniques. However, in the agricultural sector, mistakes can lead to significant financial and material losses; hence, the advice must be exact and correct. To develop an effective and trustworthy crop forecast model, a great deal of research is being done.

A system that gives Indian farmers access to predictive data is necessary if they are to make educated planting decisions. In light of this, we provide an intelligent system that, before advising the user on the best crop to grow, takes into account weather conditions (such as temperature, rainfall, and state of habitation) and soil properties (such as pH value, type, and nutrient content). The shortcomings of the current method are mitigated by the suggested methodology. Some of the tactics incorporated in the suggested system include increasing agricultural yields, enhancing decision-making, picking efficient parameters, and using IoT to assess crops in real-time. If we want accurate crop predictions, we need to use powerful algorithms. By carefully selecting the appropriate components and providing an ML model with appropriate parameters, the most valuable data might be obtained. Farmers benefit from this generated data because it suggests the best crop to cultivate. The primary goal of the crop suggestion model is to provide recommendations for the best crop to use in a given field. The amount of crops lost can be reduced by selecting the right crops for the field. The agricultural landscape is undergoing a dramatic transformation. While data-driven technologies are ushering in a new age of precision and sustainability, knowledge and experience have always been the bedrock of prosperous farming. When it comes to increasing agricultural yields and making the most efficient use of available resources, precision agriculture stands head and shoulders above the competition.

But there are many more benefits than just increased yields. Sustainable farming methods can only be advanced with data-based recommendations. Fertilizer applications, water conservation measures, and taking soil and crop specifics into account can all lead to more profitable agricultural operations. As a result, soil health improves, environmental impacts decrease, and future generations have it better.

What follows is a more in-depth examination of the following topics covered in this chapter: the researchers' prior work, the effectiveness of data-driven decision-making, the researchers' understanding of the terrain and data collection methods, the use of machine learning models for data-driven crop recommendation, the results, discussions, and a strong conclusion.

15.2   RELATED WORKS

There are studies undertaken in the area of crop recommendation using many different ML models. The Sri Lankan land shortage means that agricultural production must be maximized. To tackle this issue, a research study suggests an automated crop selection system that utilizes Arduino microcontrollers and machine learning to assess environmental factors such as temperature, water levels, and soil characteristics [1]. After processing, this data is used to suggest the best crop for a given plot of land, encouraging effective and fruitful agricultural methods.

Choosing the correct crops is a big problem for farmers in India because they do not know what kind of soil they need. Reduced productivity is a common result of this [2, 3]. Crop recommendation systems based on machine learning are suggested as a solution in a number of research projects. For increased accuracy, one method makes use of an ensemble model that combines several techniques [2]. Another study uses information from soil testing laboratories to make crop recommendations based on certain soil properties [3]. Enhancing crop choices and raising farm productivity are the goals of both approaches.

Another study emphasizes the significance of well-informed crop choices for the prosperity of Indian farmers [4]. Poor judgments that affect livelihoods can result from using traditional techniques. The suggested approach takes this into account by taking into account variables, including geographic location, sowing season, and soil data. Through analysis of these factors, the algorithm suggests appropriate crops for every unique circumstance. The expanding use of precision agricultural techniques, which optimize management procedures based on accurate data about specific fields, is also highlighted in the article.

Another research effort [5] looks at the possibilities of integrating machine learning and the Internet of Things (IoT) in Indian agriculture. This system makes use of sensors to keep an eye on a number of soil characteristics, including pH, moisture, temperature, and nutrient levels. After analyzing this data, machine learning systems suggest appropriate crops depending on the unique soil characteristics. The system also uses a convolutional neural network to detect possible plant diseases. This all-encompassing strategy seeks to maximize crop choices, strengthen soil health, and raise agricultural productivity. Strict guidelines for product use are necessary for effective pest management, which is another critical component of agriculture. An ontology-based recommendation system for pest detection and appropriate treatment suggestions based on current data is described in Reference [6]. This approach can greatly decrease the amount of manual labor needed to maintain up-to-date pest control information.

In the past, farmers frequently chose crops and applied fertilizer based on estimates. The potential for developing intelligent recommendation systems with ontologies and machine learning is demonstrated by references [7, 8]. Based on variables including soil type, location, and historical data, one algorithm suggests appropriate crops and fertilizers [7]. In a different study, machine learning is used to evaluate the characteristics of the soil, forecast crop production, and eventually recommend the best crops for particular soil types [8]. These developments support the growing field of precision agriculture, a data-driven agricultural strategy that maximizes productivity and optimizes resource use [9]. A thorough overview of machine learning applications in agriculture, encompassing topics such as disease detection, crop production prediction, animal management, and soil analysis, may be found in [10]. This demonstrates how machine learning is becoming increasingly important in building a more productive and sustainable agricultural future.

The creation of machine learning-based crop recommendation systems is the subject of numerous studies. By combining predictions from several algorithms into an ensemble model, one system achieves an astounding 99.91% accuracy [11]. Another easy-to-use smartphone application lets farmers enter information about area and soil type while using GPS to determine location. With the most accurate model achieving 95% accuracy, machine learning algorithms analyze this data to recommend profitable crops or forecast yield for selected crops [12]. With an accuracy of 97.66%, "AgroConsultant", another intelligent system, makes crop recommendations for certain places in India based on historical data (1957 – 1987) and machine learning algorithms [13]. Furthermore, crop recommendation systems are being developed with open-source data, encouraging accessibility and cooperation among researchers [14, 15].

IoT and machine learning technologies provide a potent arsenal for tackling a range of agricultural difficulties. By utilizing these technologies in enhanced disease detection techniques, yield prediction tools and crop recommendation systems; farmers will be able to make data-driven decisions that will increase their total yields and profitability [16]. The potential for integrating machine learning and IoT for thorough agricultural health monitoring is examined in Reference [17]. Utilizing unmanned aerial vehicles (UAVs), this integrated system obtains multispectral photos of crops, which are subsequently subjected to machine learning algorithms for advanced disease diagnosis. Future sustainable and effective agriculture has a great deal of promise if research and development in these areas are not stopped. Future agriculture can be more productive, resource-efficient, and less prone to crop losses by utilizing developments in data analysis, machine learning, and sensor technology.

15.3 PROPOSED METHODOLOGY

In order to make it easy to use and appealing to new users who have never used an app or facility of this kind, we have also incorporated a Graphical User Interface (GUI). The Python Tkinter package was used to add the GUI. The main modules and libraries needed to put the programming of the machine learning-based crop prediction algorithm into practice for high accuracy in land usage and crop cultivation. The implementation process includes data collection, ML model building, data preparation and training, model testing, and reducing error and increasing accuracy. Following the division of the data into training and testing sets, three distinct algorithms are used for training and testing. The model is then finalized when the best method has been selected. Currently, this model operates by receiving input from various sources and producing the best agricultural output.

As seen in Figure 15.1, we have divided the process we have proposed for our project into multiple parts. The following are the five phases: Gathering of Datasets, Feature extraction, correlation matrix, pre-processing, application of different machine learning algorithms, recommendation system, and suggested cropping.

15.3.1 DATA COLLECTION

The dataset shown in Figures 15.2 (a) and (b) contains variables such as soil pH, temperature, humidity, potassium, nitrogen, phosphorus, and rainfall. The dataset comprises 2200 samples sourced from historical records, representing eleven different crops: rice, yellow corn, chickpeas, kidney beans, pigeon peas, moth beans,

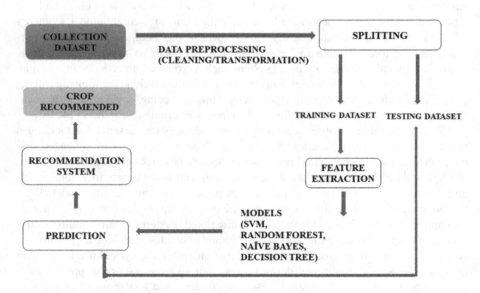

FIGURE 15.1 Block Diagram of Proposed Model.

	A	B	C	D	E
1	temperatu	humidity	ph	rainfall	label
2	20.87974	82.00274	6.502985	202.9355	rice
3	21.77046	80.31964	7.038096	226.6555	rice
4	23.00446	82.32076	7.840207	263.9642	rice
5	26.4911	80.15836	6.980401	242.864	rice
6	20.13017	81.60487	7.628473	262.7173	rice
7	23.05805	83.37012	7.073454	251.055	rice
8	22.70884	82.63941	5.700806	271.3249	rice
9	20.27774	82.89409	5.718627	241.9742	rice
10	24.51588	83.53522	6.685346	230.4462	rice
11	23.22397	83.03323	6.336254	221.2092	rice
12	26.52724	81.41754	5.386168	264.6149	rice
13	23.97898	81.45062	7.502834	250.0832	rice
14	26.8008	80.88685	5.108682	284.4365	rice
15	24.01498	82.05687	6.984354	185.2773	rice
16	25.66585	80.66385	6.94802	209.587	rice
17	24.28209	80.30026	7.042299	231.0863	rice
18	21.58712	82.78837	6.249051	276.6552	rice
19	23.79392	80.41818	6.97086	206.2612	rice
20	21.86525	80.1923	5.953933	224.555	rice
21	23.57944	83.5876	5.853932	291.2987	rice

cpdata   ⊕

	A	B	C	D	E	F
1		Crop	N	P	K	pH
2	0	Rice	80	40	40	5.5
3	1	Jowar(Sor	80	40	40	5.5
4	2	Barley(JA\	70	40	45	5.5
5	3	Maize	80	40	20	5.5
6	4	Ragi(naac	50	40	20	5.5
7	5	Chickpeas	40	60	80	5.5
8	6	French Be:	90	125	60	5
9	7	Fava bean	90	125	60	5
10	8	Lima bean	40	60	20	5
11	9	Cluster Be	25	50	25	5
12	10	Soyabean	20	60	20	5.5
13	11	Black eyec	20	60	20	5.5
14	12	Kidney be:	20	60	20	5.5
15	13	pigeon pe.	20	60	20	5.5
16	14	Moth bea	20	40	20	5.5
17	15	Mung bea	20	40	20	5.5
18	16	Green Pea	40	35	55	6
19	17	Horse Gra	20	60	20	6
20	18	Black Grar	40	60	20	5
21	19	Rapeseed	50	40	20	5

Fertilizer   ⊕

Ready   Accessibility: Unavailable

FIGURE 15.2 Dataset (a) Highlights the Climatical Values, (b) Highlights the NPK Values.

mungbean, black gram, lentil, pomegranate, banana, mango, grapes, watermelon, muskmelon, apple, orange, papaya, coconut, cotton, jute, etc.

15.3.2   DATA PRE-PROCESSING

In this stage, the null and 0 values are replaced with -1 to ensure the yield value does not affect the overall prediction. The data must be cleaned since the machine learning system cannot handle noisy, inconsistent, or incomplete input. After pre-processing, which is highlighted in Figure 15.3, we used the data set to train multiple machine learning models, including linear regression, to get the best accuracy possible.

15.3.3   FEATURE EXTRACTION

This stage aims to locate and utilize the dataset's most pertinent property. To apply classifiers, this technique eliminates redundant and unnecessary data. Numerous advantages come with feature extraction: Models can learn more efficiently and produce more accurate predictions by concentrating on the most pertinent features. (2) Machine learning algorithms demand less computer power and can train more quickly when fewer features are needed. (3) Understanding how the model generates its predictions can be simplified by concentrating on a smaller subset of significant features shown in Figures 15.4 and 15.5.

15.3.4   CORRELATION MATRIX

Once the feature has been retrieved, a correlation matrix is a useful tool for understanding their correlations. It is a visual representation of the correlation coefficient between all possible feature pairs in your dataset, as shown in Figure 15.6. When coupled, feature extraction and correlation matrix play an important role in turning unstructured data into an organized and informative manner. This enables machine learning models to draw useful information and create exact projections, resulting in more reliable crop recommendations and a brighter future for agriculture. The correlation coefficient is a statistical measure that describes the degree and direction of a linear relationship between two variables. It runs from -1 (perfect negative correlation) to +1 (perfect positive correlation), with 0 denoting no linear relationship. Advantages of using a correlation matrix: (1) The matrix aids in the identification of highly linked features, which may imply redundancy. This can help determine which features to maintain or remove during feature selection. (2) Visualizing the correlations allows you to understand how different features influence one another. This is especially important for jobs like crop recommendation, which require an understanding of the interaction between soil attributes and weather patterns. (3) The correlation matrix can help uncover potential multicollinearity concerns, which can harm model performance.

When combined, feature extraction and correlation matrix play a critical role in converting unstructured data into an organized and instructive format. This makes it possible for machine learning models to derive useful information and generate

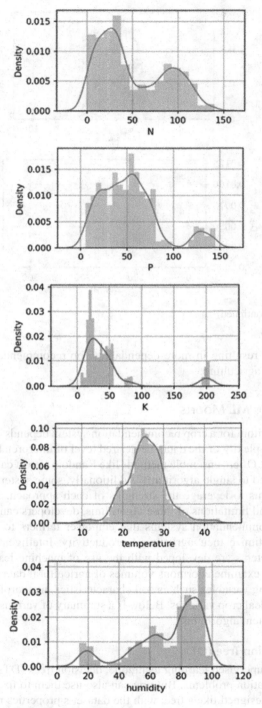

FIGURE 15.3 Analysis of Dataset wrt Crops.

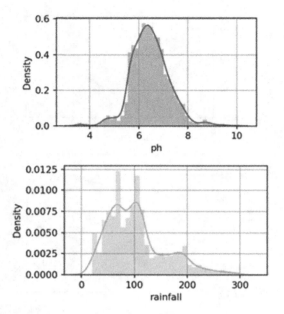

FIGURE 15.3 (Continued)

precise forecasts, resulting in more dependable crop recommendations and a more fruitful future for agriculture.

15.3.5 Various ML Models

The optimal algorithm for a crop recommendation system depends on various factors, including the complexity of the data, the desired level of interpretability, and computational resources. Often, ensemble methods like Random Forest can achieve superior accuracy compared to single algorithms. Additionally, some systems might combine multiple algorithms to leverage the strengths of each approach. By understanding the capabilities and limitations of these algorithms, developers can create robust and reliable crop recommendation systems that empower farmers to make data-driven decisions and optimize their agricultural productivity. Intelligent agricultural recommendation systems are developed with the use of machine learning techniques. These algorithms examine enormous volumes of agricultural data to find trends and connections among variables such as soil characteristics, meteorological conditions, past yields, and ideal crop varieties. Below is a summary of various popular crop recommendation system algorithms.

15.3.5.1 Decision Tree (DT)

Despite being a supervised learning technique, decision trees (DTs) are mostly used to tackle classification problems. But you can also use them to fix regression issues. The classifier is designed like a tree, with the dataset's properties represented by the

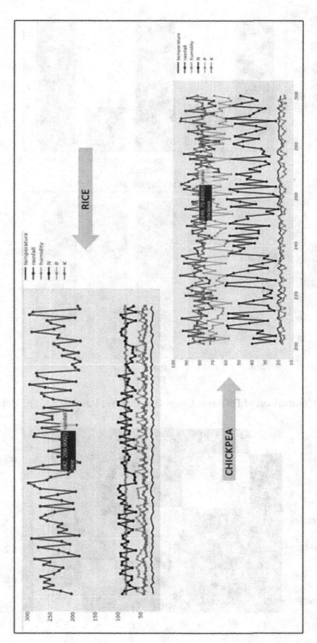

FIGURE 15.4 Different Features of Rice and Chickpea.

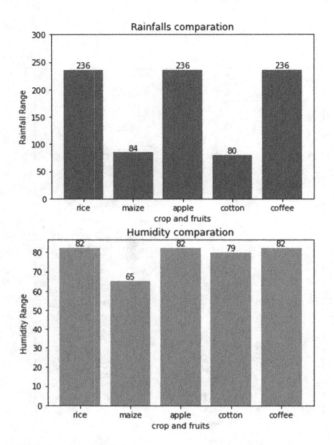

FIGURE 15.5   Comparison of Different Crops Based on (a) Rainfall and (b) Humidity.

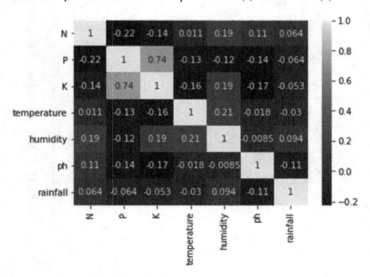

FIGURE 15.6   Correlation Among Datasets.

core nodes, the branches' decision rules, and the leaf nodes' outputs. One common application of DTs is the development of training prototypes for predicting the value or class of target variables based on decision rules learned from training data. The syntax is displayed below.

```
from sklearn.tree import DecisionTreeClassifier
```

```
DecisionTree= DecisionTreeClassifier(criterion ="entropy", random_state=2,
max_depth=5)
DecisionTree.fit(Xtrain,Ytrain)
```

15.3.5.2   Naïve Bayes (NB)

To construct classifier models, the Naïve Bayes (NB) technique assigns issue instances, represented as feature value vectors, one of a limited number of class labels. Instead of a singular method for training these classifiers, it is a set of algorithms based on a common assumption. Utilizing NB in conjunction with collaborative filtering can lead to recommendation system creation, spam filtering, real-time prediction, and the likelihood of several classes of the target attribute. It starts with figuring out the likelihood of each attribute in the dataset, often known as the class probability. The conditional probability gives each information value's probability for each class value.

```
from sklearn.naive_bayes import GaussianNB
```

```
NaiveBayes = GaussianNB() NaiveBayes.fit(Xtrain, Ytrain)
NaiveBayes.fit(Xtrain,Ytrain)
```

15.3.5.3   Support Vector Machine (SVM)

A collection of algorithms known as support vector machines, or SVMs, evaluate data for regression and classification. It is memory efficient and iteratively depicts various classes in a single plane to reduce error. This makes it one of the greatest algorithms to apply in the event that basic linear regression's error does not go away. However, due to the possibility of class overlap, its performance suffers in large and noisy datasets. Each data point is represented as a point in n-dimensional space using the SVM method, and each feature value corresponds to a particular coordinate. The next step in the classification process is to identify the hyper-plane that best distinguishes the two classes.

```
from sklearn.svm import SVC
```

```
SVM = SVC(gamma='auto')
SVM.fit(Xtrain, Ytrain)
```

15.3.5.4   Logistic Regression (LR)

While more complex logistic function expansions are available, this one uses the simplest form to depict a dependent variable that can only take on a binary value. Logistic regression, a subset of binomial regression, is employed in regression analysis for the purpose of parameter prediction in logistic models.

from sklearn.linear_model import LogisticRegression

```
LogReg = LogisticRegression(random_state=2)
LogReg.fit(Xtrain, Ytrain)
```

15.3.5.5 Random Forest (RF)

Among the most well-known and popular supervised learning methods is the Random Forest (RF) algorithm. Rather than processing the entire dataset as a single subset, it includes multiple decision trees for distinct subsets of the data. This increases the model's prediction accuracy by multiple orders of magnitude since the model determines the final output based on the majority vote of the predictions rather than on the average of all the trees' predictions. Because it can produce findings with high accuracy in a short amount of time, it is ideal even for huge and diverse datasets.

from sklearn.ensemble import RandomForestCl assifier

```
RF = RandomForestClassifier(n_estimators=20, random_state=0)
RF.fit(Xtrain, Ytrain)
```

15.4 SIMULATION RESULTS

The classification report is essential for assessing the model's performance in agriculture, especially in machine learning-powered crop recommendation systems. It thoroughly analyses the model's performance for every crop category. It has been trained to recognize important performance measures extrapolated from the report on classification:

Precision: The percentage of the model's positive predictions that come true is represented by this metric. It responds to the following query: "How many of the crops are the best choices out of all the ones the model suggests?"

Recall: This score assesses how well the model can recognize each pertinent crop recommendation. It asks: "Out of all the possible suitable crops for a specific field, how many did the model correctly recommend?"

F1-Score: This measure offers a more thorough evaluation of the model's performance by giving a balanced perspective on recall and precision.

The classification report has many benefits, which are given below in Figure 15.7.

i.   The training data or methods can be changed by identifying the model's biases against particular crop kinds.
ii. The report assists in directing efforts to improve the model for more precise crop recommendations by identifying areas that need attention.
iii. By showing a clear picture of the model's confidence in its recommendations, the report allows farmers to make well-informed decisions.

```
DecisionTrees's Accuracy is:  94.0909090909091
                 precision    recall  f1-score   support

        apple       1.00      1.00      1.00        13
       banana       1.00      1.00      1.00        17
     blackgram      0.59      1.00      0.74        16
      chickpea      1.00      1.00      1.00        21
       coconut      1.00      0.95      0.98        21
        coffee      1.00      0.86      0.93        22
        cotton      1.00      1.00      1.00        20
        grapes      1.00      1.00      1.00        18
          jute      0.78      1.00      0.88        28
    kidneybeans     1.00      0.79      0.88        14
        lentil      0.92      1.00      0.96        23
         maize      1.00      0.71      0.83        21
         mango      1.00      1.00      1.00        26
      mothbeans     1.00      0.68      0.81        19
      mungbean      1.00      1.00      1.00        24
     muskmelon      0.96      1.00      0.98        23
        orange      1.00      1.00      1.00        29
        papaya      1.00      0.89      0.94        19
    pigeonpeas      0.90      1.00      0.95        18
   pomegranate      0.94      1.00      0.97        17
          rice      0.92      0.69      0.79        16
     watermelon     1.00      1.00      1.00        15

      accuracy                          0.94       440
     macro avg      0.95      0.94      0.94       440
  weighted avg      0.96      0.94      0.94       440


Naive Bayes's Accuracy is:  0.990909090909091
                 precision    recall  f1-score   support

        apple       1.00      1.00      1.00        13
       banana       1.00      1.00      1.00        17
     blackgram      1.00      1.00      1.00        16
      chickpea      1.00      1.00      1.00        21
       coconut      1.00      1.00      1.00        21
        coffee      1.00      1.00      1.00        22
        cotton      1.00      1.00      1.00        20
        grapes      1.00      1.00      1.00        18
          jute      0.88      1.00      0.93        28
    kidneybeans     1.00      1.00      1.00        14
        lentil      1.00      1.00      1.00        23
         maize      1.00      1.00      1.00        21
         mango      1.00      1.00      1.00        26
      mothbeans     1.00      1.00      1.00        19
      mungbean      1.00      1.00      1.00        24
     muskmelon      1.00      1.00      1.00        23
        orange      1.00      1.00      1.00        29
        papaya      1.00      1.00      1.00        19
    pigeonpeas      1.00      1.00      1.00        18
   pomegranate      1.00      1.00      1.00        17
          rice      1.00      0.75      0.86        16
     watermelon     1.00      1.00      1.00        15

      accuracy                          0.99       440
     macro avg      0.99      0.99      0.99       440
  weighted avg      0.99      0.99      0.99       440
```

FIGURE 15.7 Classification Report for Different Models (a) DT, (b) NB, (c) SVM, (d) LR, (e) RF.

```
SVM's Accuracy is:  0.9727272727272728
                precision    recall  f1-score   support

       apple       1.00      1.00      1.00        13
      banana       1.00      1.00      1.00        17
   blackgram       0.94      1.00      0.97        16
    chickpea       1.00      1.00      1.00        21
     coconut       1.00      1.00      1.00        21
      coffee       1.00      1.00      1.00        22
      cotton       0.95      1.00      0.98        20
      grapes       1.00      1.00      1.00        18
        jute       0.82      0.82      0.82        28
  kidneybeans       1.00      1.00      1.00        14
      lentil       1.00      1.00      1.00        23
       maize       1.00      0.95      0.98        21
       mango       1.00      1.00      1.00        26
   mothbeans       1.00      0.95      0.97        19
    mungbean       1.00      1.00      1.00        24
   muskmelon       1.00      1.00      1.00        23
      orange       1.00      1.00      1.00        29
      papaya       0.95      1.00      0.97        19
  pigeonpeas       1.00      1.00      1.00        18
 pomegranate       1.00      1.00      1.00        17
        rice       0.73      0.69      0.71        16
  watermelon       1.00      1.00      1.00        15

    accuracy                           0.97       440
   macro avg       0.97      0.97      0.97       440
weighted avg       0.97      0.97      0.97       440
```

```
Logistic Regression's Accuracy is:  0.9568181818181818
                precision    recall  f1-score   support

       apple       1.00      1.00      1.00        13
      banana       1.00      1.00      1.00        17
   blackgram       0.94      1.00      0.97        16
    chickpea       1.00      1.00      1.00        21
     coconut       1.00      0.95      0.98        21
      coffee       0.96      1.00      0.98        22
      cotton       0.95      0.90      0.92        20
      grapes       1.00      1.00      1.00        18
        jute       0.85      0.79      0.81        28
  kidneybeans       0.93      1.00      0.97        14
      lentil       1.00      1.00      1.00        23
       maize       0.91      0.95      0.93        21
       mango       1.00      0.92      0.96        26
   mothbeans       1.00      0.95      0.97        19
    mungbean       1.00      1.00      1.00        24
   muskmelon       1.00      1.00      1.00        23
      orange       1.00      1.00      1.00        29
      papaya       0.79      1.00      0.88        19
  pigeonpeas       1.00      0.94      0.97        18
 pomegranate       0.94      1.00      0.97        17
        rice       0.79      0.69      0.73        16
  watermelon       1.00      1.00      1.00        15

    accuracy                           0.96       440
   macro avg       0.96      0.96      0.96       440
weighted avg       0.96      0.96      0.96       440
```

FIGURE 15.7   (Continued)

```
RF's Accuracy is:  0.9886363636363636
                precision    recall   f1-score    support

       apple      1.00       1.00       1.00          13
      banana      1.00       1.00       1.00          17
   blackgram      0.89       1.00       0.94          16
    chickpea      1.00       1.00       1.00          21
     coconut      1.00       1.00       1.00          21
      coffee      1.00       1.00       1.00          22
      cotton      1.00       1.00       1.00          20
      grapes      1.00       1.00       1.00          18
        jute      0.90       1.00       0.95          28
  kidneybeans     1.00       1.00       1.00          14
      lentil      1.00       1.00       1.00          23
       maize      1.00       1.00       1.00          21
       mango      1.00       1.00       1.00          26
    mothbeans     1.00       0.89       0.94          19
    mungbean      1.00       1.00       1.00          24
   muskmelon      1.00       1.00       1.00          23
      orange      1.00       1.00       1.00          29
      papaya      1.00       1.00       1.00          19
   pigeonpeas     1.00       1.00       1.00          18
  pomegranate     1.00       1.00       1.00          17
        rice      1.00       0.81       0.90          16
  watermelon      1.00       1.00       1.00          15

    accuracy                            0.99         440
   macro avg      0.99       0.99       0.99         440
weighted avg      0.99       0.99       0.99         440
```

FIGURE 15.7 (Continued)

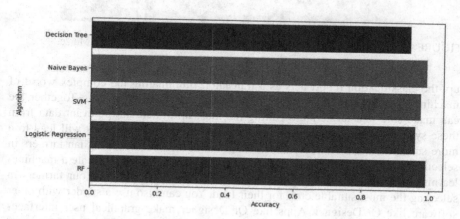

FIGURE 15.8 Accuracy Comparison.

This study has suggested several Indian crops utilizing several machine learning techniques such as Decision Tree, Naïve Bayes, Support Vector Machine (SVM), Logistic Regression, and RF. After examining these five categories of machine learning algorithms, the most accurate result was obtained by Naïve Bayes. To sum

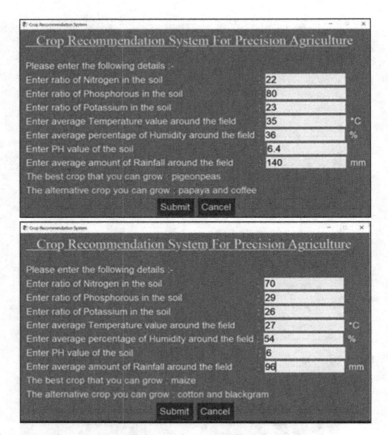

FIGURE 15.9   Output Suggesting Suitable Crop (a) for Pigeonpeas, (b) for Maize.

up, the classification report acts as a translator, illuminating the complex world of machine learning predictions for agricultural recommendation systems. Together, we can make sure that farmers are getting the most up-to-date and relevant data from these systems by fully understanding the metrics they provide. This will lead to a more sustainable and fruitful future for agriculture. To assist our Indian farmers in selecting the most productive crops for their fields, we intend to create a machine-learning program. We much appreciate your assistance in assisting our farmers in selecting the most suitable crop for their land. You can also use a spyder with other software like Ot Designer. Apps like Qt Designer make graphical user interfaces (GUIs) possible. You can build interactive GUIs with the help of the PyQt5 package, which is part of Anaconda. All of the GUI features you add will be stored in the database. For crop prediction, this is fed into the model. The crop recommendation interface gets the output from the trained model. The next step is for the GUI to recommend an appropriate crop. Knowing what to expect from their crops is a huge help for farmers. Therefore, the farmers make exact choices based on the expected result.

Initially, testing and training datasets are created using the data gathered from the public domain. The machine learning model uses a training dataset for the crop suggestion prediction model. The model receives the test data once it has been generated with the least error and the highest level of accuracy. The model that is generated receives the inputs. Next, with an accuracy of roughly 96.89%, the model forecasts and recommends when to sow the crops.

15.5 CONCLUSION

To sum up, agriculture continues to be the major pillar of our prosperity. Recent worldwide food crises, however, highlight the weaknesses in our food supply systems and the demand for a more robust agricultural system. This chapter promotes machine learning as a game-changing technology that gives farmers access to data-driven agricultural suggestions. This chapter does more than just list crops. It gives farmers the skills and resources they need to become data-savvy travelers in a world where agricultural data abounds. Precision agriculture, a data-driven strategy that maximizes resource efficiency and reduces environmental impact, is now a reality. Envision a farmer with insights about weather trends from historical records and in-the-moment observation. Thanks to this knowledge, they can choose crops with confidence and comprehensive analyses of soil health and market trends. With crops that are precisely matched to the unique characteristics of their land, this leads to maximized yields. However, the advantages go far beyond increased output. Farmers may promote sustainable practices that safeguard our valuable natural resources by minimizing excessive water usage and fertilizer application by understanding the demands of their crops.

In the end, this data-driven strategy strengthens our agriculture industry. It clears the path for abundant harvests in terms of quality and sustainability of our food production and quantity. Tradition and conjecture are no longer constraints on farmers. They have the knowledge and skills to make wise choices to protect our food system for future generations and guarantee food security. We have created and put into operation an advanced crop recommendation system that farmers across India can readily use. This approach helps farmers choose the most suitable crop to grow by considering elements such as temperature, rainfall, humidity, pH level, nitrogen, phosphorus, and potassium. Implementing the research results can enhance national productivity and generate revenue through this approach. Consequently, farmers can cultivate suitable crops, increasing both their yield and the nation's overall productivity. Our upcoming research will involve integrating yield prediction and developing an enriched dataset with numerous features.

REFERENCES

[1] Bandara, P., T. Weerasooriya, T. Ruchirawya, W. Nanayakkara, M. Dimantha, and M. Pabasara. "Crop recommendation system." *International Journal of Computer Applications* 975 (2020): 8887.
[2] Pudumalar, S., E. Ramanujam, R. Harine Rajashree, C. Kavya, T. Kiruthika, and J. Nisha (2017). "Crop recommendation system for precision agriculture." In *2016 Eighth International Conference on Advanced Computing (ICoAC)* (pp. 32–36). IEEE.

[3] Rajak, R. K., A. Pawar, M. Pendke, P. Shinde, S. Rathod, and A. Devare. "Crop recommendation system to maximize crop yield using machine learning technique." *International Research Journal of Engineering and Technology* 4, no. 12 (2017): 950–953.

[4] Priyadharshini, A., S. Chakraborty, A. Kumar, and O. R. Pooniwala (2021). "Intelligent crop recommendation system using machine learning." In *2021 5th International Conference on Computing Methodologies and Communication (ICCMC)* (pp. 843–848). IEEE.

[5] Gosai, D., C. Raval, R. Nayak, H. Jayswal, and A. Patel. "Crop recommendation system using machine learning." *International Journal of Scientific Research in Computer Science, Engineering and Information Technology* 7, no. 3 (2021): 558–569.

[6] Lacasta, J., F. J. Lopez-Pellicer, B. Espejo-García, J. Nogueras-Iso, and F. J. Zarazaga-Soria. "Agricultural recommendation system for crop protection." *Computers and Electronics in Agriculture* 152 (2018): 82–89.

[7] Chougule, A., V. K . Jha, and D. Mukhopadhyay (2019). "Crop suitability and fertilizers recommendation using data mining techniques." In *Progress in Advanced Computing and Intelligent Engineering: Proceedings of ICACIE 2017*, Volume 2 (pp. 205–213). Springer Singapore.

[8] Yadav, J., S. Chopra, and M. Vijayalakshmi. "Soil analysis and crop fertility prediction using machine learning." *Machine Learning* 8, no. 03 (2021): 41–49.

[9] Kalimuthu, M., P. Vaishnavi, and M. Kishore (2020). "Crop prediction using machine learning." In *2020 Third International Conference on Smart Systems and Inventive Technology (ICSSIT)* (pp. 926–932). IEEE.

[10] Sharma, A., A. Jain, P. Gupta, and V. Chowdary. "Machine learning applications for precision agriculture: A comprehensive review." *IEEE Access* 9 (2020): 4843–4873.

[11] Vincent, D. R., N. Deepa, D. Elavarasan, K. Srinivasan, S. H. Chauhdary, and C. Iwendi. "Sensors driven AI-based agriculture recommendation model for assessing land suitability." *Sensors* 19, no. 17 (2019): 3667.

[12] Pande, S. M., P. K. Ramesh, A. Anmol, B. R. Aishwarya, K. Rohilla, and K. Shaurya (2021). "Crop recommender system using machine learning approach." In *2021 5th International Conference on Computing Methodologies and Communication (ICCMC)* (pp. 1066–1071). IEEE.

[13] Doshi, Z., Nadkarni, S., Agrawal, R., and Shah, N. (2018, August). "AgroConsultant: Intelligent crop recommendation system using machine learning algorithms." In 2018 Fourth International Conference on Computing Communication Control and Automation (ICCUBEA) (pp. 1–6). IEEE.

[14] Jena, S. P., A. K. Yadav, D. Gupta, and B. K. Paikaray (2023). "Prediction of stock price using machine learning techniques." In *2023 IEEE 2nd International Conference on Industrial Electronics: Developments & Applications (ICIDeA)* (pp. 169–174). IEEE.

[15] Kulkarni, N. H., Srinivasan, G. N., Sagar, B. M., and Cauvery, N. K. (2018, December). "Improving crop productivity through a crop recommendation system using ensembling technique." In 2018 3rd International Conference on Computational Systems and Information Technology for Sustainable Solutions (CSITSS) (pp. 114–119). IEEE.

[16] Jena, S. P., B. K. Paikaray, J. Pramanik, R. Thapa, and A. K. Samal. "Classifications on wine informatics using PCA, LDA, and supervised machine learning techniques." *International Journal of Work Innovation* 4, no. 1 (2023): 58–73.

[17] Shukla, R., G. Dubey, P. Malik, N. Sindhwani, R. Anand, A. Dahiya, and V. Yadav. "Detecting crop health using machine learning techniques in smart agriculture system." *Journal of Scientific & Industrial Research* 80, no. 08 (2021): 699–706.

16 Application of Machine Learning in the Analysis and Prediction of Animal Disease

Soumen Nayak, Lambodar Jena, Pranati Palai, Sushruta Mishra, and Manas Kumar Swain

16.1 INTRODUCTION

The health and well-being of animals are intricately intertwined with sustainable agriculture, livestock, and industry, making a significant impact on global health and security. The precise and timely diagnosis of animal diseases is crucial not only for the profitability of farms but also for ensuring the overall welfare of animals.

Traditional diagnostic methods, rooted in clinical examination and laboratory testing, have long been the foundation of veterinary medicine. Despite their proven value, these methods have inherent limitations that are becoming increasingly evident in our ever-evolving world. A persistent challenge in animal health is the timely diagnosis, often requiring experienced veterinarians and well-equipped laboratories, making them impractical in remote or underserved areas. Moreover, clinical symptoms in animals often manifest only in the advanced stages of diseases, complicating early intervention.

In the era of data-driven technology, the integration of advanced techniques like computer vision and machine learning into veterinary diagnostics provides a promising solution. This research paper delves into the emerging field of animal disease prediction, aiming to revolutionize disease diagnosis within livestock breeding and animal husbandry.

By leveraging data-driven approaches, this initiative seeks to address the limitations of conventional methods. It involves utilizing data from various sources, including images and symptoms and harnesses contemporary technologies such as Convolutional Neural Networks (CNN) and frontend applications. CNNs, well-known for their applications in computer vision, have emerged as a revolutionary technology, and their integration into animal disease prediction is no exception. Tailored to process and analyze visual data, CNNs prove invaluable for tasks like image recognition and classification. The research explores the potential of this system to identify disease patterns before clinical symptoms appear, enabling timely intervention and

DOI: 10.1201/9781003484608-16

acting as a defense against disease outbreaks, thereby reducing economic losses in agriculture and livestock industries.

Crucially, this approach plays a pivotal role in identifying zoonotic diseases in animals, preventing their transmission to humans, and upholding public health. This introduction establishes the groundwork for exploring a comprehensive approach to animal disease prediction, emphasizing the urgency of embracing technological advancements to address the challenges posed by animal health management in the 21st century. The convergence of traditional veterinary practices with cutting-edge technologies not only promises to enhance diagnostic capabilities but also aims to redefine the landscape of animal health and welfare.

In recent years, the application of machine learning (ML) across various domains has garnered significant attention, with a notable focus on predicting and managing animal diseases. Conventional methods for disease prediction and diagnosis in animals rely on manual observations and retrospective analysis. However, the integration of ML techniques has brought about a paradigm shift in the field, offering more accurate and timely predictions. This chapter delves into the manifold applications of ML in anticipating and managing animal diseases, underscoring its potential to bolster early detection, optimize treatment strategies, and ultimately elevate animal health standards.

16.2 LITERATURE SURVEY

The potential of artificial intelligence (AI) in public health, particularly within veterinary contexts, has been recently examined by Schwalbe and Wahl [1]. They outlined four categories of AI-driven health interventions, including diagnosis, mortality and morbidity risk assessment, disease outbreak prediction and surveillance, and health policy and planning.

Recent contributions of ML to animal and veterinary public health align broadly with these categories, demonstrating versatility. Similar to medical applications in human healthcare, signal processing methods combined with ML enhance diagnostic or classification systems in animals or herds. For instance, CNNs have successfully recognized and quantified specific lesions in pigs during routine slaughtering [2].

ML's impact extends beyond imaging data. Classification tree analysis has shown improvements in enhancing the sensitivity of the classification regime for bovine tuberculosis eradication in the UK [3]. Decision Trees, a method of supervised learning, are crucial components widely used in ML, including in algorithms like random forests.

In the domain of predicting animal conditions, ML has been applied to forecast lameness in dairy cows based on milk production and conformation traits [4]. Though the predictive performance in some cases has been suboptimal, it emphasizes the potential for improvement by expanding the spectrum of training data.

In the context of health policy and planning, ML has not been extensively used for resource allocation in animal disease surveillance, as observed in public health [5]. However, ML has contributed to generating information supporting animal health surveillance planning and outbreak response. For example, ML has been applied to predict poultry population data in the USA using supervised algorithms and aerial imagery [6].

The application of ML in animal health is diverse, extending beyond the identified AI-driven intervention categories. Unsupervised ML methods have been used to discover underlying structures in poultry condemnation data and classify cattle herd types for disease control [7, 8, 9].

Moreover, ML algorithms are gaining traction in syndromic surveillance, extracting information from clinical records and automating the mining of free-text data in clinical and post-mortem reports [10, 11, 12, 13]. At the farm level, precision technologies provide large datasets for analysis, and ML algorithms are instrumental in syndromic surveillance [14, 15].

The paper [16] uses CNNs for animal disease diagnosis, achieving an impressive 97.06% accuracy, surpassing Support Vector Machine results. The crucial role of ML in predicting and detecting diseases transmissible from animals to humans underscores its significance in early identification and surveillance [17]. This paper explores ML algorithms' effectiveness in forecasting Lumpy Skin Disease Virus infection in cattle based on meteorological and geological features, highlighting the superior performance of the Artificial Neural Networks (ANN) algorithm [18]. In this study [19], ML techniques are applied for continuous animal health monitoring, demonstrating the efficacy of the Support Vector Machine with an accuracy exceeding 90%. The paper [20] introduces KATZMDA, a computational method for predicting miRNA-disease associations, showcasing its efficiency in identifying disease-related miRNAs. The paper [21] proposes a comprehensive framework for the early diagnosis of Bovine Respiratory Disease, combining ML and precision Internet of Things (IoT) technologies, outperforming existing solutions. Finally, the study [22] presents a novel SEIR-SEI-EnKF model for estimating and forecasting Dengue outbreak dynamics, contributing to proactive measures for mosquito-borne viral infection control in tropical regions.

ML algorithms showcase their versatility in addressing a growing range of tasks in animal and veterinary public health. While aligned with broad categories in global health, ML applications in veterinary contexts continue to evolve, demonstrating the expanding potential of this technology.

16.3 BACKGROUND

16.3.1 DATA ACQUISITION AND PREPROCESSING

The foundation of any successful ML model lies in the quality of the data it is trained on, and predicting animal diseases is no exception. The process of data acquisition and preprocessing is crucial, involving the careful sourcing and preparation of information from diverse channels, such as veterinary records, environmental data, and satellite imagery.

The spectrum of data types in predicting animal diseases demands meticulous preprocessing to ensure compatibility and reliability. Cleaning and standardizing data are initial steps in this journey, involving the identification and rectification of errors, outliers, or inconsistencies within the dataset. This ensures that the data is consistent and accurate, laying the groundwork for robust ML models.

Handling missing values is another critical aspect of data preprocessing. Incomplete or missing data points can significantly impact the performance of a model. Various strategies, such as imputation or removal of incomplete records, are employed to address this challenge, ensuring a comprehensive dataset for training and evaluation.

Addressing imbalances in the dataset is equally important, especially when dealing with rare or underrepresented diseases. Imbalanced datasets can lead to biased models, where the algorithm may struggle to accurately predict less common diseases. Techniques like oversampling, undersampling, or the use of advanced sampling methods help in creating a balanced representation, enhancing the model's ability to generalize across different disease categories.

Feature engineering is a pivotal step in the preprocessing pipeline. This involves selecting and transforming relevant features that contribute to the predictive power of the model. In the context of animal disease prediction, incorporating environmental factors such as temperature, humidity, and geographical location can significantly improve the accuracy of the model. These features provide contextual information that enriches the understanding of the complex interplay between environmental variables and animal health.

Moreover, advancements in remote sensing technologies and satellite imagery open up new possibilities for data acquisition. Integrating spatial and temporal information from satellite observations allows for a more comprehensive analysis of environmental factors influencing animal health. This wealth of data contributes to a holistic understanding of disease dynamics, aiding in the development of accurate and robust ML models.

In essence, the success of ML models in predicting animal diseases is intricately tied to the meticulous process of data acquisition and preprocessing. By addressing issues such as data quality, missing values, imbalances, and incorporating relevant features, researchers and practitioners ensure that the models are not only accurate but also capable of handling the complexities inherent in the prediction of animal diseases. This foundational phase sets the stage for subsequent model training, testing, and deployment, ultimately contributing to more effective disease management strategies in the field of veterinary medicine.

16.3.2 SUPERVISED LEARNING FOR DISEASE PREDICTION

Supervised learning stands as a cornerstone in predicting animal diseases, leveraging labeled datasets that encompass examples of both healthy and diseased animals. This approach involves training models to recognize patterns and associations within the data, enabling accurate predictions when faced with new, unseen cases. Classification algorithms, such as Support Vector Machines (SVM), Random Forest, and Neural Networks, have proven to be particularly effective in this domain.

SVM excels in separating classes by identifying a hyperplane that maximizes the margin between them. Random Forest, an ensemble learning method, combines multiple decision trees to enhance predictive accuracy and robustness. Neural Networks, inspired by the human brain's neural architecture, excel in capturing intricate patterns within large and complex datasets.

A notable application of supervised learning in animal disease prediction is in anticipating zoonotic diseases. Zoonoses are diseases that can be transmitted from animals to humans, and early detection in animals is crucial for preventing potential outbreaks. ML models, trained on diverse datasets encompassing health records, environmental factors, and historical zoonotic events, can analyze patterns to identify potential outbreaks.

For instance, in the context of zoonotic diseases like avian influenza or Ebola, supervised learning models can analyze patterns in animal health data to detect anomalies indicative of an impending outbreak. This proactive approach enables veterinarians and public health officials to implement preventive measures swiftly, potentially averting the spread of diseases to humans. The ability to predict and manage zoonotic diseases not only safeguards animal health but also plays a critical role in protecting public health.

The strength of supervised learning lies in its ability to generalize from labeled examples, allowing the model to make predictions on new, unseen cases. As the volume and diversity of available data continue to grow, supervised learning algorithms in animal disease prediction will likely become even more sophisticated and accurate. The ongoing integration of advanced technologies and interdisciplinary collaboration between data scientists, veterinarians, and public health professionals will further enhance the capabilities of supervised learning in the realm of animal health management.

16.3.3 Unsupervised Learning for Anomaly Detection

In the realm of predicting and managing animal diseases, unsupervised learning techniques, particularly anomaly detection algorithms, play a crucial role in identifying unusual patterns within vast and complex datasets. Unlike supervised learning, where the model is trained on labeled data, unsupervised learning operates on unlabeled data, making it particularly useful for scenarios where labeled examples are scarce or hard to obtain.

Anomaly detection algorithms within unsupervised learning are instrumental in flagging deviations from normal patterns in animal health data. Clustering algorithms, such as K-Means or hierarchical clustering, prove invaluable in grouping animals with similar health profiles. This clustering allows veterinarians to identify outliers – those animals exhibiting health patterns significantly different from the norm. Such anomalies could be indicative of the early stages of a disease outbreak, prompting swift intervention to contain and manage the situation.

Moreover, the utility of unsupervised learning extends beyond anomaly detection. These techniques are pivotal in identifying novel diseases or variations in disease presentations that may not have been previously recognized. By analyzing data without predefined labels, unsupervised learning models can unveil patterns or trends that might have otherwise gone unnoticed. This capability facilitates a more comprehensive understanding of animal health, aiding in the discovery of emerging diseases or variations in disease manifestations.

The power of unsupervised learning lies in its ability to uncover hidden patterns and anomalies without relying on predefined labels. This flexibility is particularly advantageous in the dynamic and evolving landscape of animal health, where emerging diseases and changing environmental factors can pose unforeseen challenges. The continuous monitoring and analysis of animal health data using unsupervised learning techniques contribute to a proactive and comprehensive approach to disease prediction and management.

As technology continues to advance, unsupervised learning algorithms are likely to become more sophisticated, enabling even finer detection of anomalies and novel disease patterns. The integration of these techniques with other data-driven technologies, such as real-time monitoring through IoT devices, will further enhance their capabilities in predicting and managing animal diseases. The collaboration between data scientists, veterinarians, and researchers is pivotal in harnessing the potential of unsupervised learning for the benefit of animal health and welfare.

16.3.4 REAL-TIME MONITORING AND IoT INTEGRATION

The integration of IoT devices marks a significant advancement in the capabilities of ML for predicting and managing animal diseases. This synergy between IoT and ML transforms the conventional approach to animal health monitoring by enabling real-time data acquisition and analysis.

Wearable sensors, RFID tags, and various other IoT devices play a pivotal role in this integration, providing continuous monitoring of key physiological parameters, feeding behavior, and movement patterns of animals. The real-time data generated by these devices offer a wealth of information that can be harnessed to develop more accurate and responsive disease prediction models.

ML models, particularly those utilizing anomaly detection algorithms, thrive on the constant stream of real-time data. By monitoring subtle changes in physiological patterns, feeding behaviors, or movement dynamics, these models can detect early indicators of disease symptoms. For example, a cow's rumination patterns, temperature fluctuations, or changes in gait can be monitored in real-time.

Anomaly detection algorithms are designed to recognize deviations from established norms. In the context of real-time monitoring, these algorithms play a crucial role in raising alerts when patterns deviate from the expected baseline. If, for instance, a cow displays irregular rumination patterns or an abnormal increase in body temperature, the anomaly detection system triggers an alert. This immediate notification allows for prompt intervention by veterinarians, reducing the risk of disease spread within the herd and minimizing the potential economic impact on agriculture and livestock industries.

The combination of real-time monitoring and IoT integration not only enhances the speed of disease detection but also facilitates a more nuanced understanding of animal health. Continuous data collection allows for the identification of subtle, early-stage symptoms that may go unnoticed with traditional periodic examinations.

Furthermore, this approach supports a proactive rather than reactive stance in animal health management. Veterinarians and farmers can implement timely interventions, administer targeted treatments, and implement preventive measures, ultimately contributing to improved animal welfare and reduced economic losses.

As the field of IoT continues to evolve, the integration of ML models with real-time data monitoring will likely become even more sophisticated. The collaborative efforts of technology developers, veterinary professionals, and data scientists will be instrumental in refining these systems, ensuring their effectiveness in addressing the dynamic challenges of predicting and managing animal diseases in real-world agricultural settings.

16.4 METHODOLOGY

Stage 1: Data Collection and Integration

- Gather a diverse range of data from multiple sources, including images and symptom records, ensuring a comprehensive dataset.
- Organize and centralize the collected data into a repository for efficient access and utilization in subsequent stages.

Stage 2: Data Preprocessing

- Cleanse and refine the raw data to eliminate inconsistencies, errors, or outliers, ensuring the reliability of the dataset.
- Standardize the format of images and categorize symptoms to create a uniform foundation for analysis.

Stage 3: Convolutional Neural Networks (CNN)

- Develop specialized CNN models designed to extract intricate features from images, revealing visual patterns associated with various diseases.
- Train the CNN models using extensive and diverse datasets to enhance their ability to recognize complex patterns in animal health images.

Stage 4: ML Algorithms

- Implement ML algorithms tailored for disease prediction based on symptom data.
- Train these algorithms using historical symptom data, enabling the models to learn and predict disease patterns.

Stage 5: Feature Extraction

- Extract distinct features from images using the trained CNNs, capturing visual nuances that may indicate specific diseases.

- Extract features from symptoms using ML algorithms, considering various factors contributing to disease prediction.
- Integrate the features extracted from both images and symptoms to create a more robust and comprehensive disease prediction model.

Stage 6: Model Selection and Integration

- Select appropriate models based on the characteristics of the data and the specific types of diseases under consideration.
- Seamlessly integrate the chosen models into a unified system, ensuring a synergistic approach to disease prediction.

Stage 7: Disease Prediction and Diagnosis

- Employ the integrated models to predict diseases in real-time, leveraging both image and symptom data for a more accurate analysis.
- Generate diagnostic outcomes based on the combined analysis, providing veterinarians with valuable insights for timely intervention.

Stage 8: User Interface

- Develop a user-friendly interface that simplifies the input of data and visualizes the prediction results.
- Enable users, including veterinarians and other stakeholders, to interact effortlessly with the system, fostering effective utilization.

Stage 9: Database and Storage

- Establish a secure database to store historical data, including images, symptoms, and corresponding outcomes.
- Ensure the secure and organized storage of data, supporting ongoing research and system improvement.

Stage 10: Performance Monitoring and Feedback Loop

- Continuously monitor the system's performance, assessing accuracy and reliability to maintain high standards.
- Collect user feedback and real-world outcomes, creating a feedback loop for ongoing system refinement and improvement based on practical experiences.

16.5   PROPOSED MODEL

The architecture of the proposed model is shown in Figure 16.1. It consists of the following components.

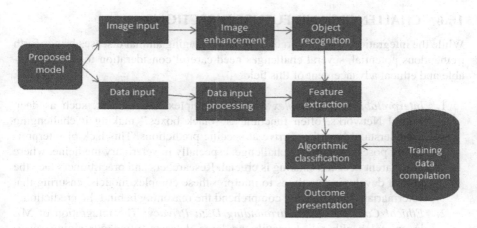

FIGURE 16.1 Architecture for the proposed ML model.

Image Input: This module of the system accepts visual input in the form of images depicting animals, which may include photographs capturing external symptoms like skin lesions, abnormal postures, or other observable indicators of potential diseases.

Data Input: In addition to images, the system requires diverse input data, covering essential information such as the animal's age, breed, medical history, observed signs, blood test results, temperature, and other pertinent health metrics.

Image Enhancement: The process of image enhancement involves manipulating images to improve their quality or extract relevant information. The objective is to prepare the images for a more thorough analysis, enhancing the visibility of symptoms or isolating specific areas of interest.

Object Recognition: A pivotal phase within image processing where the system endeavors to recognize and categorize distinctive objects within the images. For animal disease diagnosis, this may involve identifying specific anatomical features, such as the skin or regions displaying symptoms.

Feature Extraction: Subsequent to image processing, the system extracts key features from both the images and accompanying data. Features represent individual measurable properties or characteristics of the observed phenomena.

Algorithmic Classification: This stage employs algorithms capable of learning from data and making informed decisions or predictions. The system may leverage various algorithms such as decision trees, Neural Networks, SVM, and CNN based on the complexity of the task and the nature of the data.

Training Data Compilation: The training dataset serves as a curated collection of data used to train the ML model. It includes examples of various conditions with known outcomes, enabling the model to establish associations between features and outcomes for accurate predictions on new, unseen data.

Outcome Presentation: In the final step, the system presents the predicted disease based on comprehensive analysis. This presentation can manifest as a detailed report, visual representation, or an alert, providing actionable insights for timely intervention and decision-making.

16.6   CHALLENGES AND FUTURE DIRECTIONS

While the integration of ML in predicting and managing animal diseases brings forth tremendous potential, several challenges need careful consideration for the sustainable and ethical advancement of this field.

1. *Interpretability of Complex Models:* Complex ML models, such as deep Neural Networks, often function as "black boxes," making it challenging to understand how they arrive at specific predictions. This lack of interpretability poses a significant challenge, especially in veterinary medicine, where transparent decision-making is crucial. Researchers and practitioners face the task of developing methods to interpret these complex models, ensuring that veterinarians can trust and comprehend the reasoning behind the predictions.
2. *Ethical Considerations Surrounding Data Privacy:* The integration of ML in animal health relies heavily on large datasets, often containing sensitive information about individual animals, farms, and veterinary practices. Ensuring the privacy and security of this data is paramount. Addressing ethical concerns involves implementing robust data anonymization techniques, secure storage practices, and establishing clear guidelines for responsible data usage. Striking a balance between data-driven insights and individual privacy rights is crucial for fostering trust in the use of ML in animal health.
3. *Standardized Datasets:* The availability of standardized datasets is essential for training and evaluating ML models consistently. However, in the field of animal health, obtaining such datasets can be challenging due to the diverse nature of animals, diseases, and environmental factors. Efforts to establish standardized data collection protocols and collaborative initiatives to create comprehensive datasets will be instrumental in advancing the reliability and generalizability of ML models in predicting animal diseases.

16.6.1   FUTURE DIRECTIONS

1. *Integration with Blockchain for Secure Data Sharing:* Blockchain technology, known for its decentralized and secure nature, holds promise in addressing data privacy concerns. By employing blockchain for secure data sharing, stakeholders in animal health, including farmers, veterinarians, and researchers, can collaboratively contribute to and access a secure, transparent, and tamper-resistant ledger of animal health data. This not only enhances data integrity but also establishes a trustworthy foundation for ML applications in predicting and managing animal diseases.
2. *Explainable AI for Model Interpretability:* The push for "Explainable AI" (XAI) involves developing ML models that provide clear explanations of their decision-making processes. This is particularly important in contexts where decisions impact the health and well-being of animals. By incorporating XAI techniques, researchers aim to make ML models more transparent, interpretable, and accountable. This not only addresses the interpretability challenge but also fosters trust among veterinarians, farmers, and other stakeholders.

3. *Collaborative Efforts:* Collaboration between researchers, veterinarians, data scientists, and policymakers is integral to the continued success of ML applications in animal health. Multidisciplinary collaboration ensures that the technology aligns with the practical needs of veterinary professionals, adheres to ethical standards, and contributes meaningfully to both animal welfare and public health. Joint initiatives can lead to the development of robust, ethical, and universally applicable solutions for predicting and managing animal diseases.

In navigating the challenges and embracing future directions, the field of ML in animal health stands poised to make significant contributions to disease prediction, veterinary care, and overall animal welfare. The ongoing commitment to ethical practices and collaborative innovation will shape the evolution of these technologies, ensuring their positive impact on the intricate interplay between technology, animal health, and global public well-being.

16.7 CONCLUSION

The application of ML in predicting animal diseases marks a transformative shift toward proactive and data-driven veterinary medicine. The journey from early detection to real-time monitoring has showcased the invaluable contributions of ML models in safeguarding animal health and welfare. The strides made in this domain underscore the potential for technology to redefine the landscape of veterinary care, enhancing not only diagnostic capabilities but also the overall well-being of animals.

The power of ML lies not just in its ability to predict diseases but also in its capacity to adapt to the dynamic nature of animal health. The continuous advancements in technology offer opportunities for more nuanced, accurate, and timely predictions, ultimately leading to more effective disease management strategies.

Looking ahead, the collaboration between the veterinary and data science communities emerges as a linchpin for further progress. This collaboration is essential for understanding the unique challenges of animal health, refining ML models, and implementing ethical practices. As technology continues to evolve, the need for interdisciplinary cooperation becomes even more pronounced to navigate complexities and ensure that the benefits of ML are realized in a manner that aligns with the highest standards of animal welfare.

The challenges, such as model interpretability, data privacy, and the need for standardized datasets, underscore the importance of a thoughtful and ethical approach. By addressing these challenges, the collaboration between these communities will unlock the full potential of ML in veterinary medicine.

In conclusion, the marriage of ML and veterinary medicine holds promise not only for the current generation but also for the future of animal health. As we embrace technological advancements, ethical considerations, and collaborative endeavors, we pave the way for a future where the well-being of animals is safeguarded through the proactive and data-driven lens of ML. The journey ahead involves not only refining and expanding existing methodologies but also forging new frontiers where technology becomes an even more integral part of the intricate tapestry of veterinary care.

REFERENCES

1. Schwalbe N. & Wahl B. (2020). – Artificial intelligence and the future of global health. *The Lancet*, 395 (10236), 1579–1586.
2. Bonicelli L., Trachtman A. R., Rosamilia A., Liuzzo G., Hattab J., Alcaraz E. M., Del Negro E., Vincenzi S., Capobianco Dondona A., Calderara S. & Marruchella G. (2021). – Training convolutional neural networks to score pneumonia in slaughtered pigs. *Animals*, 11 (11), 3290.
3. Romero M. P., Chang Y.-M., Brunton L. A., Parry J., Prosser A., Upton P. & Drewe J. A. (2022). – Machine learning classification methods informing the management of inconclusive reactors at bovine tuberculosis surveillance tests in England. *Preventive Veterinary Medicine*, 199, 105565.
4. Shahinfar S., Khansefid M., Haile-Mariam M. & Pryce J. E. (2021). – Machine learning approaches for the prediction of lameness in dairy cows. *Animal*, 15 (11), 100391.
5. Araújo Rosas M., Benjamin Bezerra A. F. & Duarte-Neto P. J. (2013). – Use of artificial neural networks in applying methodology for allocating health resources. *Revista de Saúde Pública*, 47 (1), 128–136.
6. Patyk K. A., McCool-Eye M. J., South D. D., Burdett C. L., Maroney S. A., Fox A., Kuiper G. & Magzamen S. (2020). – Modelling the domestic poultry population in the United States: a novel approach leveraging remote sensing and synthetic data methods. *Geospatial Health*, 15 (2), 913.
7. Buzdugan S. N., Chang Y. M., Huntington B., Rushton J., Guitian J., Alarcon P. & Blake D. P. (2020). – Identification of production chain risk factors for slaughterhouse condemnation of broiler chickens. *Preventive Veterinary Medicine*, 181, 105036.
8. Buzdugan S. N., Alarcon P., Huntington B., Rushton J., Blake D. P. & Guitian J. (2021). – Enhancing the value of meat inspection records for broiler health and welfare surveillance: longitudinal detection of relational patterns. *BMC Veterinary Research*, 17 (1), 278.
9. Brock J., Lange M., Tratalos J. A., More S. J., Graham D. A., Guelbenzu-Gonzalo M. & Thulke H.-H. (2021). – Combining expert knowledge and machine-learning to classify herd types in livestock systems. *Scientific Reports*, 11 (1), 2989.
10. Dórea F. C., Sanchez J. & Revie C. W. (2011). – Veterinary syndromic surveillance: current initiatives and potential for development. *Preventive Veterinary Medicine*, 101 (1–2), 1–17.
11. Anholt R. M., Berezowski J., Jamal I., Ribble C. & Stephen C. (2014). – Mining free-text medical records for companion animal enteric syndrome surveillance. *Preventive Veterinary Medicine*, 113 (4), 417–422
12. Arguello-Casteleiro M., Jones P. H., Robertson S., Irvine R. M., Twomey F. & Nenadic G. (2019). – Exploring the automatisation of animal health surveillance through natural language processing. In *Artificial Intelligence XXXVI* (M. Bramer & M. Petridis, eds), Vol. 11927, pp. 213–226. Springer, Cham, Switzerland.
13. Bollig N., Clarke L., Elsmo E. & Craven M. (2020). – Machine learning for syndromic surveillance using veterinary necropsy reports. *PLoS One*, 15 (2), e0228105.
14. Giordano J. O., Sitko E. M., Rial C., Pérez M. M. & Granados G. E. (2022). – Symposium review: use of multiple biological, management, and performance data

for the design of targeted reproductive management strategies for dairy cows. *Journal of Dairy Science*, 105 (5), 4669–4678.

15. Sturm V., Efrosinin D., Öhlschuster M., Gusterer E., Drillich M. & Iwersen M. (2020). – Combination of sensor data and health monitoring for early detection of subclinical ketosis in dairy cows. *Sensors (Basel)*, 20 (5), 1484.

16. Mohan A., Raju R. D. & Janarthanan P. (2019, February). – Animal disease diagnosis expert system using convolutional neural networks. In *2019 International Conference on Intelligent Sustainable Systems (ICISS)* (pp. 441–446). IEEE.

17. Rehman S. (2023). – Animal disease prediction using MLmMachine lLearning techniques. *IJRASET*, 11(6), 1441–1456.

18. Afshari Safavi E. (2022). Assessing machine learning techniques in forecasting lumpy skin disease occurrence based on meteorological and geospatial features. *Tropical Animal Health and Production*, 54 (1), 55.

19. Das S., Roy R. K. & Bezboruah T. (2024). Machine learning in animal healthcare: A comprehensive review. *International Journal of Recent Engineering Science*, 11(3), 89–93.   https://doi.org/10.14445/23497157/IJRES-V11I3P109.

20. Qu Y., Zhang H., Liang C. & Dong, X. (2017). KATZMDA: prediction of miRNA-disease associations based on KATZ model. *IEEE Access*, 6, 3943–3950.

21. Casella E., Cantor M. C., Setser M. M. W., Silvestri S. & Costa J. H. (2023). A machine learning and optimization framework for the early diagnosis of bovine respiratory disease. *IEEE Access*, 11, 71164–71179.   doi: 10.1109/ACCESS.2023.3291348.

22. Yi C., Cohnstaedt L. W. & Scoglio C. M. (2021). SEIR-SEI-EnKF: a new model for estimating and forecasting dengue outbreak dynamics. *IEEE Access*, 9, 156758–156767.

17 Transforming Indian Agriculture
A Machine Learning Approach for Informed Decision-Making and Sustainable Crop Recommendations

Smitta Ranjan Dutta, Sanata Kumar Swain, Amaresh Sahu, and Swayumjit Ray

17.1 INTRODUCTION

The emerging field of crop suggestion in agriculture is capturing public attention, as farmers often lack awareness of the most suitable crops for their farms, leading to productivity hindrances and confusion. To address this, we have expanded existing databases on crop production by creating a tailored dataset for India. This dataset incorporates crucial factors such as rainfall, humidity, temperature, pH, and season, offering a comprehensive understanding of environmental and geographic influences on crop patterns suggested by Gadge, N. G. & Mahore, R. C. [1]. Utilizing this dataset, we aim to develop a machine-learning model that can assist in determining the optimal crops for specific regions. The implementation of machine learning in agriculture holds the potential to revolutionize the industry by providing early guidance on raw material and resource requirements. This proactive approach addresses issues like nutrient shortfalls resulting from the cultivation of inappropriate crops, thereby enhancing overall output efficiency.

Recognizing India's lag in adopting modern agricultural solutions, especially given its significance as the primary source of income for a majority of the population, there is a pressing need for scientific advancements in the sector. The primary objective of our model is to furnish farmers with guidance based on variables such as soil composition, temperature, humidity, rainfall, and geographical impact.

In our research, we present an overview of the data collection and analysis procedures used to evaluate the effectiveness of machine learning algorithms in predicting crop yields for various crops. The dataset encompasses information from 22

DOI: 10.1201/9781003484608-17

different crops between 2000 and 2014. By employing machine learning methods, we train and test the model to assess its accuracy. The methodology involves data preparation using Pandas and NumPy, data visualization with Matplotlib and Seaborn, and machine learning model development using Scikit-learn, XGBoost, and CatBoost. The integration of gradient-boosting algorithms like XGBoost and CatBoost proves essential in creating precise prediction models that account for intricate interactions between variables.

17.2 LITERATURE REVIEW

In recent years, numerous studies have delved into the convergence of agriculture and advanced machine learning, shedding light on the potential applications of contemporary technology in the agricultural sector.

Kevin Tom Thomas et al. [2] specifically focused on crop prediction using K-Nearest Neighbors (KNN) with cross-validation, giving prominence to soil factors. Mahendra N. and colleagues [3] opted for the Decision Tree method to predict crops, taking into account both soil and weather factors. Some researchers, including Kevin Tom Thomas, utilized the Support Vector Machine (SVM) algorithm to determine rainfall, a critical weather parameter.

Mansi Shinde et al. [4] conducted their study on NPK (nitrogen, phosphorous, and potassium) levels in the soil for crop recommendations, employing the Random Forest algorithm. Sonal Jain et al. [5] underscored the influence of weather and soil characteristics on crop selection, noting the absence of crucial soil features like NPK levels. P. Suresh et al. [6] applied modified KNN and K-Means to predict crop yields in Tamil Nadu, concentrating on main crops.

For detecting soil type for the recommendation of crops, S. Pudumalar and co-authors [7] employed an ensemble strategy that combined Naïve Bayes, K-Nearest Neighbor, and Random Forest algorithms. R. Kumar and his research team [8] conducted an investigation where they took into account factors such as governmental policies, market prices, and production rates during the meticulous process of crop selection. The introduction of the Crop Selection Method (CSM) was proposed as a means to maximize the crop yield rate.

The integration of the Internet of Things (IoT) into crop recommendation takes center stage in the investigation led by Angu Raj and his research team [9]. They employed sensors to collect soil characteristics, including temperature, humidity, soil moisture, and pH. In a similar vein, Lakshmi N. and her research team [10] proposed a comprehensive crop recommendation system by leveraging advanced big data methodologies. Their approach involved considering a range of factors, including drainage, texture, color, depth, soil erosion, pH, permeability, and water retention.

Vivek, M.V.R., and his colleagues [11] undertook a comprehensive investigation into the application of diverse machine learning algorithms for crop recommendation. Their research was grounded in the analysis of meteorological data, crop data, and soil data, incorporating a variety of techniques such as Naïve Bayes, multilayer perceptron, JRIP, Jf48, and SVM.

17.3   DATA AND VARIABLES

Using the data gathered from kaggle.com [12], we were able to create a predictive model that will suggest the best crops to grow on a certain farm depending on a variety of factors. The size of the dataset is 2200. Nitrogen, phosphorous, and potassium are the components of soil. The following is the list of data fields:

The nitrogen content ratio is represented as N. The Phosphorous content ratio is represented as P. The Potassium content ratio is represented as K. The oC temperature is represented as temperature. Humidity in percent is represented as humidity. The soil's pH value is represented as pH. Rainfall in mm is represented as rainfall. Crop suitability is represented by a label.

The snapshot of the data set is depicted in Figure 17.1.

Figure 17.1 depicts the crop recommendation dataset, which contains information on various agricultural parameters for crop recommendation. The dataset includes features such as temperature, humidity, rainfall, and soil type, along with the corresponding recommended crop labels. It aims to assist in predicting suitable crops based on environmental conditions, aiding agricultural decision-making processes.

17.4   METHODOLOGY AND MODEL SPECIFICATIONS

The development of this research code involved the utilization of diverse technologies, including but not limited to NumPy, Pandas, and Matplotlib. Pytorch, along with several other Python libraries, was also instrumental in the coding process. The models for this project were prepared through the utilization of Python files, .ipynb files, and .pkl files. A range of machine learning algorithms was employed on the provided dataset, leading to the training and evaluation of numerous models. The

	N	P	K	temperature	humidity	ph	rainfall	label
0	90	42	43	20.879744	82.002744	6.502985	202.935536	rice
1	85	58	41	21.770462	80.319644	7.038096	226.655537	rice
2	60	55	44	23.004459	82.320763	7.840207	263.964248	rice
3	74	35	40	26.491096	80.158363	6.980401	242.864034	rice
4	78	42	42	20.130175	81.604873	7.628473	262.717340	rice
...	...	...	...	...	...	...	...	...
2195	107	34	32	26.774637	66.413269	6.780064	177.774507	coffee
2196	99	15	27	27.417112	56.636362	6.086922	127.924610	coffee
2197	118	33	30	24.131797	67.225123	6.362608	173.322839	coffee
2198	117	32	34	26.272418	52.127394	6.758793	127.175293	coffee
2199	104	18	30	23.603016	60.396475	6.779833	140.937041	coffee

FIGURE 17.1   Crop recommendation dataset.

outcomes of these models enable users to predict the optimal crop for cultivation based on specific conditions through an intranet-based system. This system facilitates the addition, viewing, and updating of crops suitable for cultivation in a particular area under specific environmental conditions, thereby enhancing agricultural management practices.

The models used in this research paper are:

I Logistic Regression

For binary classification issues in machine learning, logistic regression is a popular statistical method. Unlike linear regression, which provides predictions about continuous outcomes, logistic regression predicts the likelihood that an input will belong to a certain class. In particular, it is useful when the dependent variable is categorical and has two alternative outcomes, such as true or false, 0 or 1.

A logistic function, sometimes referred to as the sigmoid function, is used by the logistic regression model to the linear combination of input features and their associated weights. The output of this transformation is mapped to the range [0, 1], which represents probabilities. Mathematically, the logistic regression model can be represented as:

$$P(x) = \frac{1}{1+e^{-(z)}} \tag{17.1}$$

Here,

$P(x)$ represents the probability of the target variable being 1 given the input x.

The base of the natural logarithm is e, or roughly 2.71828.

$Z = b_0+b_1x_1+b_2x_2+\dots\dots+b_nx_n$ where $b_0,b_1,b_2,\dots\dots,b_n$ are the coefficients of logistic regression model.

$X_1,X_2,\dots\dots,X_n$ are the predictor variables.

II Multilayer Perceptron

Multilayer Perceptrons (MLPs) are a subset of Artificial Neural Networks (ANNs) comprised of numerous interconnected layers of nodes, similar to neurons. Due to the feedforward nature of the network, information flows from the input layer through the hidden levels and out to the output layer in a single direction.

In an MLP, non-linearity is usually introduced into the network by each neuron using an activation function. The hyperbolic tangent (tanh) function, rectified linear unit (ReLU) function, and sigmoid function are the three activation functions most frequently utilized. The network can identify intricate links and patterns in the data thanks to these activation mechanisms.

MLPs are characterized by their architecture, including the number of layers, the number of neurons in each layer, and the connectivity between layers. The first layer is the input layer, which receives the features of the input data. The last layer is the output layer, which produces the network's predictions. Any layers between the input and output layers are known as hidden layers.

During the training process, an MLP learns to map input data to output predictions by adjusting the weights and biases of the connections between neurons. This adjustment is typically done using backpropagation, a gradient-based optimization algorithm that calculates the gradients of the loss function with respect to the network parameters and updates them accordingly.

MLPs are capable of learning complex, nonlinear relationships in data and are widely used for various machine-learning tasks, including classification, regression, and pattern recognition. However, they can be prone to overfitting if not properly regularized or if the training data is insufficient. Regularization techniques such as dropout and weight decay are often employed to mitigate overfitting in MLPs.

III EXTREME GRADIENT BOOST

XGBoost, short for Extreme Gradient Boosting, is a powerful and scalable implementation of the gradient boosting framework. Designed for speed and performance, it has become one of the most popular and widely used machine learning algorithms for structured/tabular data. Gradient boosting is a complex learning technique that creates a strong predictive model by combining several weak learners that are usually. on a decision basis trees in a row XGBoost improves traditional gradient boosting algorithms by adding several innovative features: Regulation: XGBoost includes L1 and L2 regularization terms to the objective function to avoid overfitting and improve generalization. Gradient optimization: It uses a more efficient algorithm to find optimal weights during the boosting process, which speeds up training and improves accuracy. Tree Pruning: Unlike traditional gradient boosting, XGBoost greedily grows and prunes trees using a method called "leaf growth by depth". This strategy maximizes the reduction of the loss function and results in more efficient and accurate trees. Handling missing values: XGBoost has built-in functions to handle missing values in the input data, allowing it to efficiently use incomplete data sets. Cross-Validation: It supports built-in cross-validation functions to facilitate hyperparameter tuning and model performance evaluation. Parallel: XGBoost is highly parallel and can take advantage of multi-core processors to speed up training. It also supports graphics processing unit (GPU)acceleration for even higher speed. Flexibility: XGBoost supports many loss functions and evaluation metrics, making it suitable for a variety of supervised learning tasks, including regression, classification, and ranking.

IV RANDOM FOREST CLASSIFICATION

Random Forest Classifier is a machine learning algorithm that belongs to the ensemble learning family. It is built on the idea of Decision Tree classifiers. Instead of relying on a single Decision Tree, a Random Forest builds multiple decision trees and combines them together to produce a more accurate and stable prediction. Working of Random Forest:

Random sampling: a Random Forest randomly chooses a subset of the training data with replacement (this process is called bootstrapping). This subset is used to train each Decision Tree in the forest. This sampling technique ensures tree diversity.

Random feature selection: At each node in the Decision Tree, the Random Forest selects a random subset of objects instead of considering all features in the split. This helps reduce the correlation between individual trees and further increases diversity.

Tree construction: Each Decision Tree in a Random Forest is constructed independently using a randomly selected subset of data and features. Trees usually grow to maximum depth without pruning.

Voting: After all trees are built, each tree makes separate predictions. In classification tasks, each tree "votes" for the class it predicts, and the class with the most votes becomes the final prediction. In regression tasks, the final prediction is often the average of all tree predictions.

Random Forest classifications offer several advantages:

They are robust to overfitting due to the combination of many trees.
They are robust, high-dimensional data and large data sets efficiently.
They are robust to outliers and data noise.
They provide estimates of feature importance that can be useful for feature selection.

V   K-Nearest Neighbors (KNN) Classifier

Easy to understand and apply to both regression and classification problems, KNN is a machine learning technique. Instance-based, or lazy learning, algorithms are the kind that this one uses; this means that during training, no model is explicitly learned. In its place, new data points are predicted by comparing them to the training dataset, which it has learned by heart.

Training Phase: During the training phase, KNN stores all the available data points and their corresponding labels (in the case of classification) or target values (in the case of regression).

Prediction Phase: When a new data point is given, KNN identifies the K-Nearest Neighbors to this data point from the training set. "Nearest" is typically defined using a distance metric such as Euclidean distance, Manhattan distance, or cosine similarity.

Classification: For a classification task, the algorithm assigns the class label that is most frequent among the K-Nearest Neighbors. In other words, the class label with the majority vote among the neighbors is assigned to the new data point.

Regression: For a regression task, the algorithm assigns the average of the target values of the KNN as the prediction for the new data point.

The choice of the parameter k (the number of nearest neighbors to consider) is crucial in KNN. A smaller value of k may provide a more flexible decision boundary but may increase the effect of noise on the data. Conversely, a larger value of k may provide a smoother decision boundary but may cause the algorithm to ignore known local patterns.

Some key features of the KNN algorithm are:

Simple implementation: KNN is easy to understand. and deploy, making it a good choice for beginners and rapid prototyping.

Nonparametric: KNN makes no assumptions about the distribution of the underlying data, allowing it to be used efficiently on data with complex patterns.

Memory intensive: Because KNN remembers the entire training set dataset, it can be memory intensive, especially for large datasets.

Sensitive to feature scale: KNN is sensitive to feature scale, so features often need to be scaled or normalized before running the algorithm.

KNN is often used in various applications such as recommender systems, image recognition, and anomaly detection. However, its performance can be significantly degraded for high-dimensional data and large datasets due to the curse of dimensionality, making the concept of closeness less important in higher-dimensional spaces.

VI THE NAÏVE BAYES

Based on the "naïve" premise of feature independence and the Bayes theorem, Naïve Bayes is a straightforward yet effective probabilistic classification technique. Naïve Bayes, which is used for text categorization and other problems where the assumption of independence holds, is a simple algorithm that frequently works unexpectedly well in reality.

The likelihood of a hypothesis, given evidence, or $P(h \mid e)$, is described in terms of the likelihood of the evidence given the hypothesis, or $P(e \mid h)$, and the prior probability of the hypothesis, or P(h). Naïve Bayes is predicated on this theorem.

Mathematically, it is represented as:

$$P(h \mid e) = P(e \mid h) . P(h) / P(h) \tag{17.2}$$

Feature Independence: The "naïve" assumption of Naïve Bayes is that all features are independent of each other because of their class label. In other words, the presence of a given feature in a class is not related to the presence of any other feature.

Classification: Using a Bayes phrase, Naïve Bayes determines the likelihood of each class based on the case's characteristics in order to classify a new instance. The class that has the best chance of being the anticipated class is chosen. Naïve Bayes Types: Naïve Bayes classifiers come in a number of varieties, such as:

Naïve Gaussian Cells: Continuous features are assumed to have a Gaussian distribution.

Multiname Naïve Bayes: Suitable for discrete feature classification (text classification, for example, word count).

The Bernoulli Though similar to polynomial Naïve Bayes, Naïve Bayes is appropriate for binary features. The following are some benefits of Naïve Bayes classifiers: They are efficient in terms of computation and grow well in terms of both feature and training instance counts. They are adept in handling big data sets and high-dimensional data. They offer probabilistic forecasts with room for ambiguity.

Still, naïve Bayesian classifiers might not work effectively in the presence of independence. if there is a violation of the assumption or if the traits have a strong correlation. They are also inappropriate for activities that require the capture of correlations between intricate elements.

For tasks like text categorization and spam filtering, in particular, Naïve Bayes classifiers are widely used due to their ease of use, efficacy, and efficiency.

VII STOCHASTIC GRADIENT DESCENT CLASSIFIER

A stochastic gradient descent (SGD) classifier is a linear classifier that uses a stochastic gradient descent optimization algorithm to find the optimal parameters for classification.

Due to its efficiency and low data processing, it is particularly useful for large-scale machine-learning tasks.

The SGD classifier works as follows:

Initialization: The classifier starts with a first guess of the parameters (weights and biases).

Training: training data is iteratively fed into the classifier. At each iteration (or epoch), one training instance (or miniset of instances) is randomly selected (hence "stochastic" in SGD) and used to update the model parameters.

Gradient Descent: The model parameters are updated at a certain time. Direction, which minimizes the loss function (e.g., joint loss for linear for SVM, log loss for logistic regression). This update is done using the gradient of the loss function with respect to the parameters.

Regulation: Alternatively, to avoid overfitting, you can add regularization conditions to the loss function (e.g., L1 or L2 regularity). Convergence: The training process continues until a convergence criterion is met (e.g., maximum number of iterations or a small change in the loss function).

The main advantages of the SGD classifier are:

Efficiency: SGD is computationally efficient, especially for large. datasets because it updates the model parameters based on one instance or miniset at a time.

Scalability: It can handle datasets with a large number of samples and features, as well as high-dimensional sparse data tasks commonly found in text classification. and natural language processing.

Flexibility: The SGD classifier supports various loss functions (e.g., hinge loss, log loss) and can be easily extended to different types of linear models, including linear SVM, logistic regression, and linear regression.
However, SGD has some limitations:

Sensitive to feature scale: SGD is sensitive to the scale of input features, thus, feature scaling (e.g. normalization or normalization) is often necessary.

Hyperparameter tuning: the performance of the SGD classifier depends on the selection of hyperparameters, such as learning rate, regularization strength, and group size, which may require tuning to achieve optimal performance.

Convergence problems: SGD may converge to a suboptimal solution when the learning rate is too high or too low or when the data is noisy.
     In general, the SGD -classifier is a versatile and powerful algorithm that is suitable for many classification tasks, especially in scenarios where computing resources are limited or when dealing with large datasets.

VIII THE DECISION TREE CLASSIFIER

Frequently employed for both classification and regression applications, a Decision Tree is a well-liked and simple machine learning approach.
     Within this flowchart layout, a "test" property is represented by each internal node, a test result by each branch, and a class label (in classification) or numerical value (in regression) by each leaf node.

The following is how Decision Trees function:

Choose the optimal function from the data set that offers the most significant distribution of the data: this is done by the algorithm. In most cases, this is accomplished by using metrics like variance reduction in regression or the Gini additive in classification. For this node, the optimal partition is determined by taking the function and its threshold into consideration.

Partitioning: Using the function value that has been chosen, the dataset is split up into smaller groups. The best features are then selected and split recursively for each subset, and each subset is then utilized to construct child nodes, until one stopping requirement is satisfied. Breaking point: When one of the predetermined breakpoints is reached, the splitting comes to an end. fulfills certain requirements, such as: Every data point in a node is part of the same class (for classification). The tree's maximum depth attained. We have reached the minimum amount of node samples. Leaf node count reached its maximum.

Optional pruning: To lessen overfitting and increase generalization efficiency, after a tree is constructed, extra branches can be cut off using pruning procedures.

A benefit of this approach is that the algorithm uses the decision criteria of each internal node up until it reaches the leaf node in order to navigate the Decision Tree and forecast new cases. The projected identifier (or value) for the leaf node is then allocated to the case based on the class identifier (or regression value) associated with it. Decision trees offer a number of benefits:

Interpretation: Decision trees are very helpful for obtaining insight into data and elucidating model predictions since they are simple to comprehend and analyze.

Nonparametric: Because they do not assume anything about feature relationships or data distributions, decision trees are appropriate for nonlinear relationships.

Managing several kinds of data: Without feature scaling or monthly coding, decision trees are able to handle both numerical and categorical data.

Decision trees do have several drawbacks, though:

Overfitting: Decision trees are prone to overfitting training data, particularly when given room to expand. too deep or if there is noise in the data.

Instability: Different decision trees can be greatly impacted by little changes in the data, making them noise sensitive.

Multilevel features: It can produce biased results since decision trees favor features with more levels or classes.

Though they have drawbacks, decision trees and more sophisticated variations, such as gradient-boosted trees and random forests, are frequently employed in machine learning applications and serve as the foundation for numerous more intricate algorithms. They come in handy, especially when working with data that is mixed with numerical and categorical features or when interpretability and transparency are crucial.

17.5 EMPIRICAL RESULTS

Figure 17.2 represents confusion matrices for various models: Logistic Regression, Decision Tree, Random Forest, Ada Boost, Bagging, SVM (RBF & Linear Kernel), KNN, Naïve- Bayes (multinomial & Gaussian), Multilayer Perceptron, XGBoost, LightGBM, Catboost and SGD classifier.

Figure 17.2 represents a collection of 16 scatter plots arranged in a grid. Each scatter plot represents a confusion matrix of a different machine-learning model. The confusion matrix is a specific table layout that allows visualization of the performance of the model. Each plot shows four quadrants representing True Positives, True Negatives, False Positives, and False Negatives. The confusion matrices

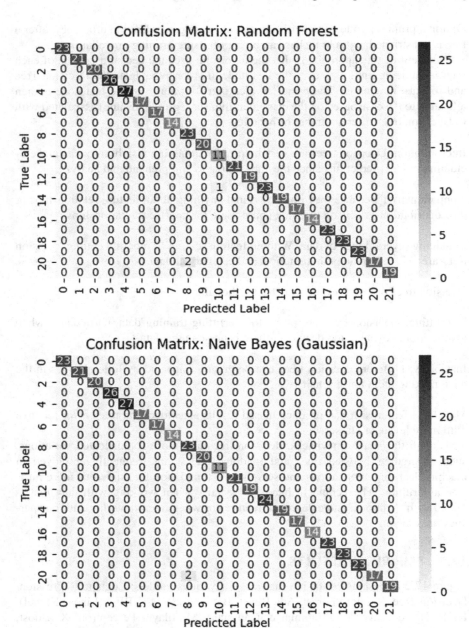

FIGURE 17.2   Confusion matrix for machine learning models.

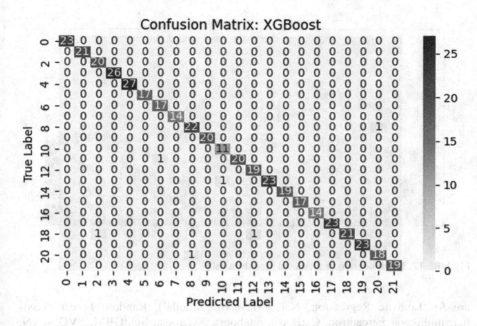

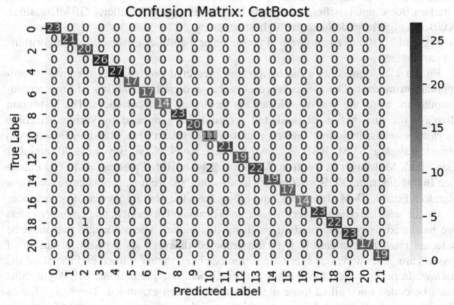

FIGURE 17.2 (Continued)

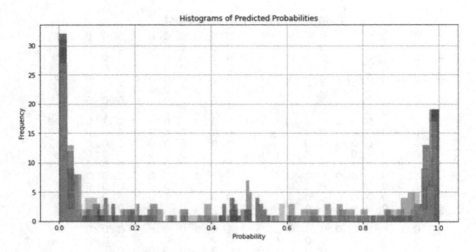

FIGURE 17.3   Histogram of predicted probabilities.

are for Logistic Regression, Naïve Bayes(multimodal), Random Forest classifier, multilayer perceptron, CatBoost, adaboost, XGBoost, lightGBM, SVC, KNN, GradientBoostingClassifier, AdaBoostClassifier, BaggingClassifier, LGBMClassifier. XGBoost algorithm delivers the highest accuracy.

Figure 17.3 shows a histogram of predicted probabilities of various ML algorithms on parameters.

Figure 17.3 represents a histogram that predicted probabilities for crop recommendation models. The x-axis represents the predicted probability of crop recommendation, while the y-axis indicates the frequency of occurrences. The histogram displays the distribution of predicted probabilities generated by various machine learning models, including logistic regression, Decision Tree, Random Forest, Ada Boost, Bagging, SVM (RBF & Linear Kernel), KNN, Naïve Bayes (Multinomial & Gaussian), Multilayer Perceptron, XGBoost, LightGBM, Catboost, and SGD classifier. In this diagram, XGBoost has provided the highest frequency of occurrence. Then Random Forest, SVM algorithms, and others have the next frequency of occurrence.

On the selected dataset, every algorithm is evaluated & trained, and various models are produced by running predictions using this model. The algorithm that must be selected must be compatible with the prepared datasets. As a result, a variety of models are generated after every algorithm has been purely trained and tested on the dataset. In order to develop models and make accurate predictions, ML Algorithms must be coded once all of those algorithms have been examined. The Naïve Bayes Extreme Gradient Boost delivers the highest and equal accuracy, or 99.55 percent, out of all the algorithms. The accuracy percentage obtained with Random Forest and SVM is 99.31. Regression using logistics and k-NN yields accuracy scores of 98.86 and 98.63, respectively. Decision trees only provided an accuracy of 85.91 percent in the case, which was their worst performance.

17.6 FUTURE SCOPE

IoT and Sensor Data Integration: Future research could explore IoT-integrated devices and sensor data into a machine learning framework. This would enable real-time monitoring of environmental conditions such as soil moisture, temperature and humidity, and provide more accurate and timely information on crop recommendations.

Adaptation to climate change: As climate change increases, agriculture will be affected. Climate modeling and predictive analytics must be incorporated into machine learning models. Researchers could explore how machine learning algorithms can adapt to changing climate conditions and provide robust crop recommendations better suited to future climate scenarios.

Plant disease detection and management: Building on the existing framework, future research could focus on integrating image processing and deep learning techniques for early detection and management of crop diseases. By analyzing images of plants and leaves, machine learning models can identify disease symptoms and recommend appropriate measures to farmers.

Adapting to regional variation: Due to differences in climate, soil and soil, agricultural practices vary considerably across regions in India. clipping settings. Future research could develop region-specific machine learning models tailored to the unique agricultural characteristics of each region. This would improve the accuracy and relevance of crop recommendations to local farmers.

Incorporation of Socio-Economic Factors: Future research could explore integrating socio-economic data into a machine learning framework to address socio-economic challenges in Indian agriculture. By considering factors such as market prices, labor availability, and farmer preferences, machine learning models can provide more comprehensive and context-aware recommendations for sustainable crop management.

Fetching services and farmer adoption: Successful implementation of machine learning based. crop recommendations require effective extension services and farmer adoption. Future research could focus on developing strategies to encourage the adoption of machine learning technologies among farmers, including training programs, knowledge platforms and policy incentives.

Assessing long-term sustainability: ensuring the long-term sustainability of agricultural practices. Future research could include assessing the environmental, economic, and social impacts of machine learning-based crop recommendations. This would include monitoring indicators such as soil health, water use, crop stability and farmers' livelihoods to assess the overall sustainability of the approach.

17.7 CONCLUSION

We utilized a sample dataset from Kaggle that contained data culled from a sizable agricultural demographic. Farmers frequently take a hit-and-miss technique that wastes resources and land or even makes crops grow uncontrollably. We are trying to remove all these heavy hurdles by providing them with access to an accurate and compelling model generated by machine learning using a Random Forest classifier to decide the correct crop to be produced in their fields. This will enable them to raise their agricultural production's quality and volume. By doing this, they will be able to maintain the nutrients and quality of the soil. Before harvesting a particular crop, farmers could run into problems or sickly crops. They are then allowed to upload the crop and soil reports. Subsequently, addressing challenges involves the exploration of viable AI-driven approaches to present practical remedies. Furthermore, the implementation of AI models opens avenues for the provision of APIs and Virtual agents, facilitating IoT solutions. This technological integration empowers farmers to engage seamlessly with suppliers of essential resources, such as seeds and fertilizers, tailored to the specific needs of each crop.

REFERENCES

[1] Gadge, N. G., & Mahore, R. C. (2022). A Novel Application of Machine Learning for Crop Analysis and Prediction. In Proceedings of the IEEE International Conference for Women in Innovation, Technology, and Entrepreneurship (ICWITE **2022**). https://doi.org//10.32628/CSEIT228681

[2] Thomas, K. T., Varsha, S., Saji, M. M., Varghese, L., & Thoma, E. J. (2020). Crop Prediction Using Machine Learning. International Journal of Future Generation Communication and Networking, 13(3), 1896–1901. http://sersc.org/journals/index.php/IJFGCN/article/view/28189

[3] Mahendra, N., Vishwakarma, D., Nischitha, K., Ashwini, & Manjuraju M. R. (2020). Crop Prediction Using Machine Learning Approaches. International Journal of Engineering Research & Technology (IJERT), 09(08). DOI:10.17577/IJERTV9IS080029

[4] Shinde, M., Ekbote, K., Ghorpade, S., Pawar, S., & Mone, S. (2016). Crop Recommendation and Fertilizer Purchase System. International Journal of Computer Science and Information Technologies, 7(2), 665–667. https://doi.org//10.1007/978-981-13-0224-4_19

[5] Jain, S., & Ramesh, D. (2020). Machine Learning Convergence for Weather-Based Crop Selection. In Proceedings of the IEEE International Students' Conference on Electrical, Electronics, and Computer Science (SCEECS) (pp. 1–6). https://doi.org//10.1109/SCEECS48394.2020.75

[6] Suresh, P., Kumar, P. G., & Ramalatha, M. (2018). Prediction of Major Crop Yields of Tamilnadu Using K-Means and Modified KNN. In Proceedings of the 3rd International Conference on Communication and Electronics Systems (ICCES) (pp. 88–93). https://doi.org//10.1109/CESYS.2018.8723956

[7] Pudumalar, S., Ramanujam, E., Rajashree, R. H., Kavya, C., Kiruthika, T., & Nisha, J. (2017). Crop Recommendation System for Precision Agriculture. In Proceedings of the Eighth International Conference on Advanced Computing (ICoAC) (pp. 32–36). https://doi.org/10.1109/ICoAC.2017.7951740

[8] Kumar, R., Singh, M. P., Kumar, P., & Singh, J. P. (2015). Crop Selection Method to Maximize Crop Yield Rate Using Machine Learning Technique. In Proceedings of the International Conference on Smart Technologies and Management for Computing, Communication, Controls, Energy and Materials (ICSTM) (pp.138–145). http://doi.org//10.1109/ICSTM.2015.7225403

[9] Raj, A., Balashanmugam, A., Dr. Thiyaneswaran, Jayanthi, J., Yoganathan, N., & Srinivasan, P. (2021). Crop Recommendation on Analyzing Soil Using Machine Learning. Turkish Journal of Computer and Mathematics Education (TURCOMAT), 12, 1784–1791. http://doi.org//10.17762/TURCOMAT.V12I6.4033

[10] Lakshmi, N., Priya, M., Shetty, S., & Manjunath, C. R. (2018). Crop Recommendation System for Precision Agriculture. *International Journal for Research in Applied Science & Engineering Technology (IJRASET)*, 6(5). http://doi.org//10.22214/ijraset.2018.5183

[11] Vivek, M. V. R., Sri Harsha, D. V. V. S., & Maran, P. S. (2019). A Survey on Crop Recommendation Using Machine Learning. *International Journal of Recent Technology and Engineering (IJRTE)*, 7(5C). www.ijrte.org/wp-content/uploads/papers/v7i5c/E10300275C19.pdf

[12] Siddharth, S. (n.d.). Crop Recommendation Dataset. Retrieved from www.kaggle.com/datasets/siddharthss/crop-recommendation-dataset

18 Automated Detection of Water Quality in Smart Cities Using Various Sampling Techniques

Sanket Mishra, T. Anithakumari, and Ojasva Jain

18.1 INTRODUCTION

By the year 2050, it is projected that 70% of the global population will be urban dwellers, as stated by various global entities, including the United Nations. The escalating urban population and the shortage of adequate resources for sustainable living underscore the importance of employing advanced technologies such as the Internet of Things (IoT) and artificial intelligence (AI) in urban planning to tackle diverse social challenges. The concept of a 'smart city' emerged in the early 1990s, and it was envisioned to leverage communication and computing technologies to build and enhance urban infrastructures. The development and application of smart cities, driven by IoT and AI, are expected to induce profound transformations worldwide. Water remains a fundamental necessity for people, yet the globe is currently grappling with the challenge of safeguarding less than 0.5% of its accessible potable water (Goparaju 2021). Approximately 1.8 billion individuals, or 28% of the world's population, are consuming unsafe drinking water (Miller et al. 2019). In many cities, there is a shortage of clean drinking water, and about 45% of rural populations are using water that does not meet safety standards. The survival of life on Earth is heavily reliant on water. All living beings need water of a reliable quality for their survival. The quality of water significantly influences public health and environmental conditions. Research by Haghiabi et al. (2018) has shown that water serves various purposes, including drinking, agriculture, and industrial processes. Arun & Premkumar (2021) have noted that many waterborne diseases in India stem from the consumption of contaminated water, leading to health issues such as fluorosis, acute diarrheal diseases, typhoid, and cholera. According to a United Nations report (Aldhyani et al., 2020), contaminated water is responsible for approximately 1.5 million deaths annually due to water-related diseases. The report further highlights that in developing nations, 80% of health issues are linked to contaminated water, leading to 2.5 billion cases of illness and five million deaths each year. The World Economic Forum has indicated that the crisis in drinking water poses a significant global risk, contributing to an infant mortality rate of approximately 200 deaths daily. The research by Koditala

DOI: 10.1201/9781003484608-18

& Pandey (2018) points out that unsafe drinking water is the main reason for about 3.4 million deaths annually. Following these findings, Vasudevan & Baskaran (2021) have emphasized the requirement for consistent observation of water quality. Shaibur et al. (2021) focus their research on the physical, chemical, and biological aspects of water quality. Kedia (2015) addresses the inefficiencies of traditional methods, which involve manual collection and laboratory testing of water samples. These methods are costly, time-consuming, and not apt for handling real-world data. The work also investigates the need for intelligent systems to monitor the water quality in the real world. Studies by Koditala et al. (2018) suggest the necessity for ongoing water quality monitoring. Research by Nasir et al. (2022) and Sahour et al. (2023) demonstrate that machine learning, through data analysis, enhances predictive capabilities in water quality monitoring. Vasudevan et al. (2021) emphasize the implementation of continuous, real-time water quality monitoring using IoT, cloud computing, and machine learning technologies.

In this work, we propose the IoT framework "Shuddh". The objective of Shuddh is to address the water contamination problem in different cities in India. To this end, we have taken the Indian water quality data set from Kaggle[1]. The dataset contains water quality parameters from various states in India. To identify the water quality, we employed different machine-learning approaches to the data. The data is balanced using different sampling techniques, and features are extracted using Principal Component Analysis (PCA). The balanced data with reduced feature sets is fed to the top-performing machine learning approach, and its performance is evaluated using standard classification metrics. This work depicts the effectiveness of the best model through an ablation study elaborated in the forthcoming sections. The next paragraph describes the organization of the paper and represents the different aspects of the work.

The rest of the paper has been organized as follows. Section 2 describes the related work done using different approaches. Section 3 briefly describes the data set, Section 4 contains the details about the sampling methods used to balance the dataset and Section 5 deals with various ML approaches implemented in the research work. Section 6 explains preprocessing and feature selection in our dataset, followed by an experimental evaluation and results. In the last section, we discuss the conclusion and future directions.

18.2   RELATED WORKS

In this section, we thoroughly review previous studies on monitoring water quality using different machine-learning techniques. Monitoring water quality is essential to protect human health, the environment, and maintain water standards. Nasir et al. (2022) applied several machine learning classifiers, including Random Forest (RF), Logistic Regression (LR), Decision Tree (DT), CATBoost, Extreme Gradient Boosting (XG-Boost), and Multilayer Perceptron (MLP) to categorize water quality data using the Water Quality Index (WQI). The results indicated that the CATBoost model was the most precise, achieving an accuracy of 94.51%.

Shafi et al. (2018) developed a real-time embedded system to measure water quality parameters across different sources in Pakistan. This study indicates that previous water quality assessments were conducted primarily manually rather than by automated means. To forecast water quality, several machine learning techniques, such as Support Vector Machine (SVM), K-Nearest Neighbors (k-NN), and deep neural networks, were used. Of these, the deep neural networks achieved the highest performance, with an accuracy of 93%.

The study by Radhakrishnan & Pillai (2020) analyses the detection of water quality using different machine learning models, including SVM, DT, and Naïve Bayes. Using four key aspects of water quality, including pH, dissolved oxygen (DO), electrical conductivity, and biochemical oxygen demand (BOD) of the water, the models are examined and validated. Two real-time datasets are subjected to an intensive simulation study, and the results of three machine-learning techniques are reported. Based on the findings, the DT method is determined to be the most appropriate classification model.

The research by Asadollah et al. (2021) discusses how to predict water quality by considering chemical and physical parameters as input variables. Based on the correlation analysis, the combination of variables is determined, and machine learning models are used to predict the water and identify the best predictive models. The results indicate that the extratree regression model performs better in predicting WQI compared to other models.

AI techniques such as the nonlinear autoregressive neural network (NARNET) and Long Short-Term Memory (LSTM) models were employed to forecast the WQI. It was found that the NARNET model outperformed the LSTM models in terms of the R-square achieved. Furthermore, algorithms such as SVM, k-NN, and Naive Bayes were used to categorize the WQI data, the SVM algorithm recording the highest accuracy at 97.01% compared to k-NN and Naive Bayes.

The study by Abuzir & Abuzir (2022) demonstrated the application of machine learning methods for the analysis of water quality. The multilayer perceptron (MLP) model was found to outperform other models in terms of accuracy. To further enhance this accuracy, the PCA dimensionality reduction technique was employed, which isolates important and pertinent features from the data set.

Abyaneh (2014) carried out a study to evaluate the effectiveness of ANN and MLP models in forecasting water quality. The research utilized BOD and COD parameters at a wastewater treatment facility in Iran. The models' effectiveness was assessed based on the coefficient of correlation (r) and the root mean square error (RMSE). It was found that the ANN model outperformed the MLP model, achieving an accuracy of 79%.

Research by Khan et al. (2022) showed that machine learning algorithms are effective in forecasting and categorizing water quality using different parameters such as pH, DO, SS, EC, turbidity, chloride, COD, TDS, and alkalinity. During data preprocessing, the median technique was employed to manage missing values, and the min-max scaler was utilized for data normalization. The principal component regression (PCR) technique was then used for prediction. The findings indicate that PCA combined with Support Vector Regression (SVR) outperforms other methods, achieving an accuracy of 95%. Efforts were also made to minimize PCA components. Subsequently, the best outcomes were observed with multiple linear regression (MLR). A gradient-boosting classifier was used to classify the status of water quality and compared against various AdaBoost classifiers, a support vector classifier, and an RF classifier. The gradient boosting classifier proved to be the most effective and accurate, achieving a classification accuracy of 100%.

The work identifies 13 characteristics from 20 stations to predict water quality in the Karuna River in Iran using various ML algorithms such as MLR, M5P, RBF-SVR, and RFR.MLR, RFR, and RBF-SVR show the best precision in the 20 stations to estimate the parameters TDS, SAR, and TH, but M5P gave unsatisfactory results in all 20 stations. Input variables affect the performance of the model, so in order to improve and select effective features better, a PCA dimension reduction approach was used. ML algorithms are good at predicting water quality, so they reduce time and cost to some extent.

Khullar & Singh (2021) assessed the water quality in the Yamuna River, classifying it into categories such as excellent, poor, or unfit for consumption. They developed a novel hybrid machine-learning technique that employs an ensemble learning approach. This technique achieved a high accuracy rate of 99. 65%, demonstrating its superiority in predicting water quality compared to other methods such as naive Bayes, SVM, bagged trees, and boosted trees. Zamri et al. (2022) aim to predict the water quality of a river in Malaysia, which presents a challenge due to its vast data set with diverse characteristics that complicate the assessment of pollution levels. The KNN method is used to determine the most precise classifications of river quality, taking into account various distances and weights. In terms of distance, the accuracy of KNN is evaluated against SVM and DT. KNN enhanced with entropy shows superior performance, achieving an accuracy of 99.90%. Azrour et al. (2022) developed a model using three different machine learning algorithms to predict WQI and another three to classify water quality. The measured parameters were temperature, pH, turbidity, and coliform. The results indicated that the MLR method provided the most precise WQI predictions. Thus, it outperforms gradient boosting and Lasso regression. Furthermore, using ANN for water quality classification resulted in an accuracy of 85.11% and a precision of 89.01%, making it more effective than SVM or decision trees.

Khan et al. (2022) utilized machine learning approaches to develop a predictive model addressing both the WQI and its classification. The model included four essential water parameters: temperature, pH, turbidity, and coliforms. Employing multiple regression algorithms was key, resulting in an accuracy of 85.11% for predicting the WQI. This section examines different studies that evaluate water quality using various machine-learning techniques. The subsequent section will present a novel framework named "Shuddh" and will delve into its components thoroughly.

18.3   PROPOSED FRAMEWORK

The following section describes the various components or modules of the proposed framework depicted in Figure 18.1.

18.3.1   DATASET DESCRIPTION

The data set in this study encompasses the quality of drinking water and its attributes gathered from multiple states in India. This data, collected over a period of nine years, comprises 1679 entries, of which about 1100 were preserved post-data preprocessing. This selection was necessary due to the presence of noise, missing values, and discrepancies in the collected data Jaloree et al. (2014).

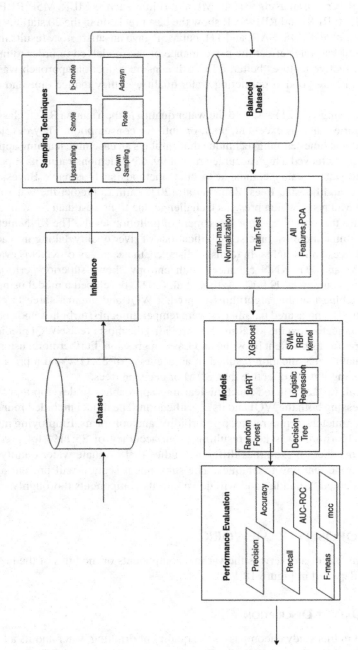

FIGURE 18.1 Shuddh framework.

18.3.2 DATA PREPROCESSING

Enhancing data quality is essential during the data preprocessing stage in data analysis. This phase is crucial for removing irrelevant records and purifying the data prior to its utilization in models Aldhyani et al. (2020) and Nasir et al. (2022). The presence of missing values can greatly compromise the performance of the classification model and can result in skewed results, which could invalidate the findings. Therefore, addressing missing values is imperative as stated by Zahin et al. (2018). Furthermore, Donders et al. (2006) indicate that the use of a single imputation approach may lead to an overestimation of precision and uncertainty in estimating missing values. An advanced technique, multiple imputation using chained equations (mice), provides a more precise estimation of the actual results Madley-Dowd et al. (2019). An online tutorial by Alice (2018) suggests that a missing data threshold of 5% is acceptable for large data sets.

The data has been displayed annually to pinpoint the year with the highest amount of missing data, which will then be excluded. According to Figure 18.2, the years

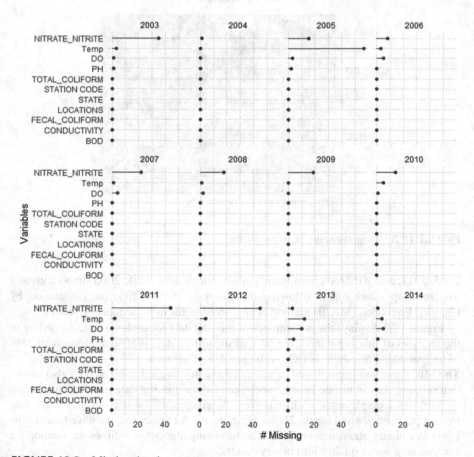

FIGURE 18.2 Missing data in years.

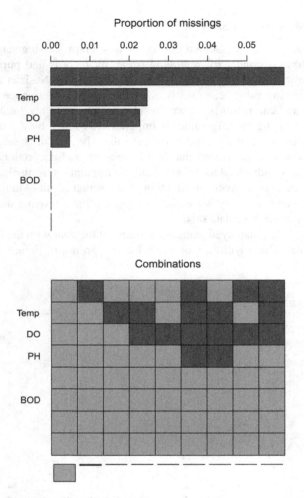

FIGURE 18.3 Proportionality of missing data.

2003, 2011, and 2012 have significant gaps in nitrate data, while 2005 shows a greater loss in temperature data. Following this analysis, the Multivariate Imputation by Chained Equations (MICE) method is employed to fill in these missing values.

Figure 18.3 displays the percentages of missing data for each variable according to the analysis of missing data. Over 50% of the nitrate data is absent, along with 25% of temperature data, 20% of DO levels, and 5% of pH data.

The MICE method is employed to address these missing values. The data also shows that missing data in nitrate is often accompanied by missing data in other variables.

Table 18.1 displays the multi-category classification of water samples according to their percentages and labels. However, in the model presented, we have treated this study as a binary classification issue. The following algorithm outlines the method for categorizing water quality into binary results.

Algorithm 1 Binary Classification of WQI

1: Input: WQI
2: **if** WQI ≤ 25% **then**
3:      Output: Clean
4: else
5:      Output: Unclean
6: end if

18.3.3 CALCULATION OF WATER QUALITY INDEX

The computation of the WQI involves averaging the individual index values of selected physio-chemical parameters Das Kangabam et al. (2017). This average is used to determine the WQI. In this research, water samples were classified based on their WQI scores, and the WQI was derived using the equation below:

Utilizing the physiological and chemical properties of water, the researcher Das Kangabam et al. (2017) assessed WQI and classified the water quality in Table 18.1 according to their percentages. A WQI range from 0% to 25% is defined as clean, 26% to 50% as unclean, 51% to 75% as polluted, and 76% to 100% as extremely contaminated.

$$WQI = \frac{\sum_{j=1}^{m} \left(q_j x w_j \right)}{\sum_{j=1}^{m} \left(w_j \right)} \qquad (18.1)$$

We have computed the pH and DO values by using the Eq. (18.2).

$$q_j = 100 \times \left[\frac{\left(V_j - V_{ideal} \right)}{\left(S_j - V_{ideal} \right)} \right] \qquad (18.2)$$

$$w_j = \frac{K}{S_j} \qquad (18.3)$$

TABLE 18.1
Multi-Class WQI Classification

WQI Percentages	Categories
0- 25%	clean
26 – 50%	unclean
51 – 75 %	polluted
76 - 100%	highly polluted

where,

m represents a number of parameters, q_j is calculated for each parameter quality estimate scale, and w_j denotes the Unit weight of the parameter, s_j and v_j are used to measure tested water sample values, and Videal is the ideal value of the feature in clean water.

18.3.4 DATA SAMPLING TECHNIQUES

The water potability data set used in this study was obtained from Kaggle Indian water quality data. Upon analyzing the data, it was evident that the class labels were not evenly distributed. Consequently, various data sampling methods were implemented to balance the classes. In practical scenarios, especially in datasets gathered from various sectors of a smart city, class imbalances often occur. Such imbalances arise when one class has significantly fewer samples than others, leading to biased predictions that favor the class with more samples and neglect the minority class. This imbalance adversely impacts the classification performance, affecting both the accuracy and precision of the data. Furthermore, in binary classification, it was observed that one class termed the minority class, had significantly fewer instances than the other, known as the majority class. Datasets with imbalances can lead to skewed predictions and inaccuracies in class prediction rates Tyagi & Mittal, (2020). Techniques at the data level that address these imbalances strive to create an even distribution by adjusting the ratios of instances from both the majority and minority classes. Resampling is used as a technique to mitigate the imbalance by modifying the number of samples from each class.

Figure 18.4 shows the imbalance of the data set, which has 33 clean water samples and 1059 unclean samples.

Figure 18.4 illustrates the imbalanced dataset of WQI, wherein the volume of unclean water samples significantly surpasses that of clean water samples. This disproportion may compromise the accuracy of class predictions and yield results that are disproportionately biased toward unclean water. To mitigate such discrepancies, specific sampling techniques have been employed to equilibrate the dataset and enhance the precision of the predictive outcomes.

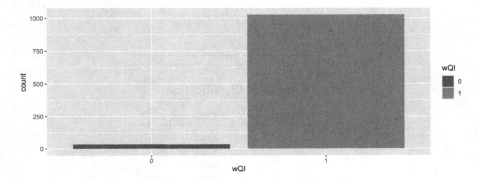

FIGURE 18.4 Class distribution of water quality dataset.

18.3.5 Undersampling

The undersampling technique is particularly effective for large datasets. It enhances performance in terms of both speed and storage by reducing the number of training samples Tyagi & Mittal (2020). This method involves removing less critical patterns through heuristic rules or random selection. According to researcher Vida (2016), some common undersampling strategies include the condensed nearest neighbor rule and one-sided selection. However, there is a risk associated with undersampling as it might overlook important data.

Figure 18.4 illustrates the disparity in the dataset, which initially contains 33 clean and 1059 unclean samples. To equalize the data set through undersampling, the number of unclean samples is reduced to 770, matching the 770 clean samples. The balanced dataset is then depicted in Figure 18.5.

Figure 18.5 illustrates the uneven distribution of data in scenarios involving clean and unclean water. The subsequent section will explore the need for sampling methods and address the issues arising from this data imbalance.

18.3.6 Oversampling

Oversampling is used when dealing with a small and unbalanced data set. It achieves balance by duplicating entries from the underrepresented class. The benefit of this approach is the preservation of all potential data, while its drawback is the introduction of duplicate observations to the dataset, which can lead to model overfitting. Researchers have used various oversampling techniques, such as the Synthetic Minority Oversampling Technique (SMOTE) and Probability Density Function

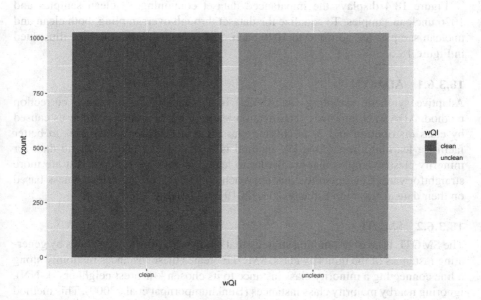

FIGURE 18.5   Balanced data with undersampling.

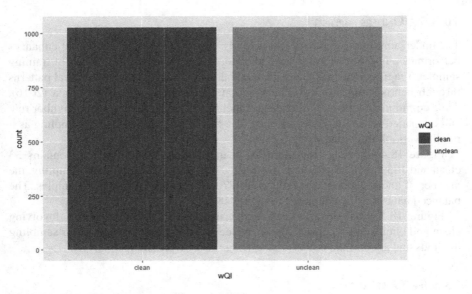

FIGURE 18.6   Balanced data with oversampling.

Estimation Based Oversampling (PDFOS) Tyagi & Mittal (2020). In scenarios of class imbalance, common machine learning algorithms often show a bias toward the majority class, considering the minority class as a negligible error. Consequently, this bias prevents the model from recognizing unseen instances of the minority class Li et al. (2022).

Figure 18.4 displays the imbalanced dataset containing 33 clean samples and 1026 unclean samples. To equalize the dataset through oversampling, both clean and unclean samples are adjusted to 1026 each. The balanced dataset is then illustrated in Figure 18.6.

18.3.6.1   ADASYN

Adaptive synthetic sampling (ADASYN) is a binary class imbalance correction method. ADASYN improves classification learning by 1) minimizing biases caused by class distribution and 2) modifying classification decision boundaries to better facilitate learning. The ADASYN method is designed to create synthetic data for minority class examples that are harder to learn, as opposed to those that are more straightforward. Consequently, this approach enhances learning effectiveness based on their data distribution Elreedy et al. (2019) and He et al. (2008).

18.3.6.2   SMOTE

The SMOTE is an oversampling strategy that operates in overlapping zones by generating instances of the minority class. SMOTE creates these instances randomly along a line connecting a minority class instance to its chosen k-nearest neighbors (k-NN), ignoring nearby majority class instances (Bunkhumpornpat et al., 2009). This method enhances the representation of the positive class to boost the importance of certain areas within the feature space. SMOTE synthesizes new instances in the feature space through random sampling along the line that links the instance and its k-nearest neighbors instead of simply duplicating existing data (k-NN). Although SMOTE is broadly recognized in scholarly circles, it is not without its flaws, including tendencies toward overgeneralization and increased variance (Vida, 2016).

18.3.6.3 Borderline-SMOTE

The authors, as referenced in Han et al. (2005), have conducted a review indicating that most classification algorithms strive to accurately define the borderline of each class during the training process to improve the quality of the prediction. The study shows that examples distant from the borderline contribute less to prediction accuracy, whereas those on or near the borderline are more likely to be misclassified. While SMOTE creates new synthetic examples along the line connecting a minority example to its nearest neighbors, the Borderline-SMOTE method focuses exclusively on oversampling or reinforcing the examples on the borderline (Chawla et al., 2002). The Borderline-SMOTE algorithm is designed to produce positive synthetic instances. According to Tyagi et al. (2020) and Han et al. (2005), the algorithm categorizes minority instances into three groups: SAFE, DANGER, and NOISE, based on the number of neighbors with the majority. Only instances classified as DANGER, where the number of majority neighbors exceeds the minority, are targeted for synthetic instance generation to address potential misclassification.

18.3.6.4 ROSE

The Random Oversampling Examples (ROSE) technique, introduced by Lunardon et al. (2014), creates artificially balanced samples through a smoothed bootstrap process, aiding in both model estimation and assessment stages. ROSE mitigates the impact of class imbalance. The samples produced by ROSE are regarded as providing a more accurate representation of the original data Menardi et al. (2014). In terms of assessing learner accuracy, ROSE can utilize different methods, including holdout, bootstrap, and cross-validation. However, Lunardon et al. (2014) have noted that holdout and bootstrap methods tend to overestimate accuracy.

PDFOS, an oversampling methodology, generates synthetic data by approximating a Gaussian multivariate distribution. This approach involves estimating and extrapolating the probability density function of the existing minority class samples to produce congruent instances. The balanced data is depicted in Figure 18.7.

Algorithm 2 Oversampling PDFOS on WQI Dataset

1: Require $R = \{y_j = (w_1^{(j)}, \ldots, w_d^{(j)})\}_{j=1}^n$
2: Require N, number of positive instances required to balance
3: Initialize $R' = \varnothing$
4: Search for h which minimizes M(h)
5: Find V, unbiased covariance matrix of R

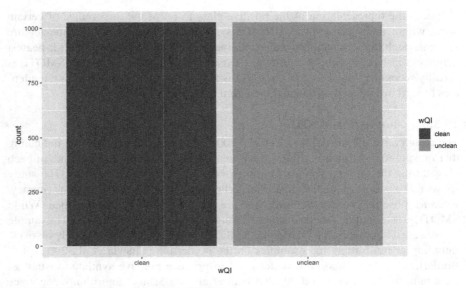

FIGURE 18.7   Balanced data with PDFOS.

6: Calculate $V = SS^T$ by Cholesky decomposition
7: for $j = 1$ to N do
8:    Choose $y \in R$
9:    Pick r with respect to normal distribution $r \sim N^d(0, 1)$
10:    $R' = R' \cup \{y + hrS\}$
11: end for
12: Return R', which yields synthetic positive instances

18.4 NORMALIZATION

Normalization constitutes a pivotal data preparation methodology widely utilized in machine learning. This technique scales the data to a uniform range from 0 to 1. In particular, Mustaffa et al. (2011) have explored a variety of feature scaling methods, including Min-Max normalization Raju et al. (2020), Z-Score normalization Henderi et al. (2021), and Decimal Scaling Singh & Singh (2020). In the context of this research, Min-Max normalization is employed for data normalization. The Min-Max normalization method applies a linear transformation to the original data, effectively maintaining the intrinsic relationships among the values of the data. This technique is crucial for fitting the data within a normalized range of 0 to 1. The calculation of Min-Max normalization is articulated through the equation Han et al. (2022):

$$y_n = \left| \frac{\left(y_0 - y_{min}\right)}{\left(y_{max} - y_{min}\right)} \right| \tag{18.4}$$

where, Yn is the normalized data
ymin represents the minimum values of the data
ymax represents the maximum values of the data
where yo is a value taken from the dataset.

18.5 SPLITTING TRAINING AND TESTSET

In this study 'dplyr' package used to split data into Training & Test Set in 75% and 25% split. The complete dataset has 1059 observations. After splitting training set has 795 observations and test set has 265 observations. Figure 18.8 shows the kernel density plot of the training and test set distribution in the dataset.

Figure 18.9 illustrates the relationship between the features in the data set. The proposed data set has seven features: total coliform, temperature, dissolved oxygen,

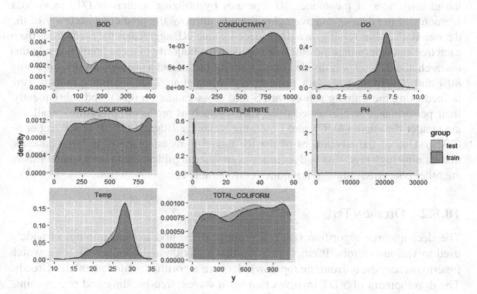

FIGURE 18.8 Kernel Density plot for Training &Test set distribution.

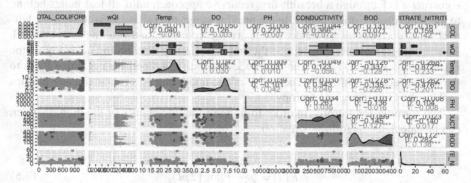

FIGURE 18.9 Pearson correlation with features in dataset.

pH, conductivity, BOD, and nitrate. The Pearson correlation is the statistical metric that shows the linear relationship and direction between the features in the data set. The relationship can be positive or negative. The positive relationship signifies that the variation

In the following section, we discuss the various machine learning approaches that are implemented on the data set to predict water quality.

18.6 MACHINE LEARNING APPROACHES

18.6.1 RANDOM FOREST

Random Forest (RF) is a widely recognized algorithm in machine learning and falls under the category of supervised learning. It is applicable to machine learning problems that require both classification and regression tasks. The foundation of RF is ensemble learning, which combines multiple classifiers to solve complex problems and improve the effectiveness of the model. RF operates by utilizing numerous DTs in various segments of the dataset and averaging them to improve the predictive accuracy of the dataset. Rather than relying on a single decision tree, RF aggregates the predictions from each tree and determines the outcome based on the majority of these predictions. In this research, the data set was divided into a training subset and a testing subset, comprising 80% and 20% of the total data, respectively (Wang et al., 2021). Initially, models were trained using the training subset with 5-fold cross-validation (CV), and subsequently, their performance was assessed using metrics such as precision, precision, recall, the area under the curve (AUC), and f-measure Rasaei & Bogaert (2019) and Wang et al. (2021). One of the advantages of the RF model is its adaptability to large datasets, maintaining high precision and accuracy even when significant data is missing, or there are other concerns, although it requires considerable computation time.

18.6.2 DECISION TREE

The decision tree algorithm (DT), a prominent method in data mining, is widely used in various sectors. It employs a top-down, divide-and-conquer strategy, which functions recursively from the top down. Its core algorithm is fundamentally Greedy. The development of a DT involves two main stages: tree-building and tree pruning. Initially, during the tree-building stage, a portion of the training data is selected to construct a DT through a breadth-first recursive approach until all leaf nodes belong to the same class. Subsequently, in the pruning stage, the DT is assessed with the leftover data, corrections are made, nodes are incorporated, and pruning is conducted. This pruning phase aims to reduce the impact of noisy data on the accuracy of the classification in the DT construction process, which is a repetitive cycle that leads to the final structure of the DT. The purity of the nodes can be evaluated using metrics such as the Gini index and entropy.

The formula for entropy, in order to find out the uncertainty or the high disorder, can be calculated as follows equation:

$$E(s) = \sum_{i=1}^{k} -Pllog(2pl) \qquad (18.5)$$

where, 'p', denotes the probability of entropy and E(S) denotes the entropy. DT necessitates a low level of initial processing, yet they are more effective for classification tasks compared to regression tasks. Training them is costly, and they exhibit significant time complexity Khullar & Singh (2021).

18.6.3 Extreme Gradient Boosting

Extreme Gradient Boosting (XGBoost) is a centralized learning model based on gradient-boosting decision trees that effectively addresses the problem of overfitting and improves predictive accuracy, making it a significant area of study in the computer science field. This algorithm is widely utilized in classification and prediction tasks as part of the boosting technique. The advantages of XGBoost include rapid processing speeds, high model accuracy, and straightforward usability. It demonstrates superior computational performance relative to other fundamental algorithms Chen & Guestrin (2016). The advantages of XGBoost identified by Asselman et al. (2021) include its ability to manage large data sets and prevent overfitting in clean data scenarios. However, it faces challenges such as being hard to interpret in the presence of noisy data and being susceptible to overfitting.

18.6.4 Logistic Regression

Logistic regression (LR) is a type of supervised learning algorithm that addresses classification challenges where the dependent variable is either dichotomous or categorical. The primary application of LR is in binary logistic regression, which deals with binary outcomes (yes or no). In logistic regression, the outcomes are discrete and confined to specific values. LR is employed to predict the likelihood of an event occurring Krhoda & Amimo (2019). mathematical equation to calculate the probability of an event:

$$h(x) = \frac{1}{1 + e - (B0 + B1x)} \quad (18.6)$$

is the independent variable. (B0+B1X) is derived from equation of line Y (predicted) = (B0 + B1X) + Error V value

Tu (1996) explores the advantages and disadvantages of LR. It is quick to train and straightforward. However, LR struggles with multi-class classification and performs poorly in the presence of correlated attributes in the dataset.

18.6.5 Support Vector Machine

SVM is a supervised machine learning method suitable for both classification and regression tasks. Among the top-performing ready-to-use supervised algorithms is SVM. It functions as a binary linear classifier. The primary objective of such a classifier is to establish a hyperplane that effectively segregates data points. SVM is capable of consistently predicting time series data, even when the underlying system behaviors are nonlinear, non-stationary, and undefined beforehand Muharemi et al. (2019). As a widely recognized approach, the SVM classifier employs kernel techniques to project data into a higher-dimensional space Ladjal et al. (2020). The observed data set can be represented as follows:

$$(p_i, q_i), (p_i - 1), i = 1, \ldots k \tag{18.7}$$

where k represents the number of observations, p denotes the distribution in space, and q is mentioned as the corresponding class label or output. To determine the optimal separating hyperplane used the vector V and a constant c under the various conditions. The main idea of SVM is to minimize the margins of the hyperplane in order to achieve good efficiency.
Where

$$(\vec{v}.\vec{x} + c \geq 0)$$

$$if q_i = +1, (\vec{v}.\vec{x} + c \geq +1), if q_i = -1, (\vec{v}.\vec{x} + c \leq -1) \tag{18.8}$$

This SVM model offers advantages in that it can still generate favorable results even with limited data availability Yu et al. (2004). Additionally, studies highlight some drawbacks, such as the increased time required to train the model with large datasets and the impact of personal circumstances and variable weights on the model's effectiveness.

18.6.6 BAYESIAN ADDITIVE REGRESSION TREES

Sparapani et al. (2021) presented a user-friendly version of the Bayesian additive regression trees (BART) through the BART package. This package stands out as a leading Bayesian non-parametric, tree-based ensemble method in machine learning. Consistent with machine learning standards, BART identifies the relationships between covariates, x, and the response variable, giving f (x) while allowing flexibility in not requiring the user to define the functional form of f or interaction terms among covariates. Additionally, BART supports variable selection, which is particularly advantageous in high-dimensional scenarios, facili-tated by an optional sparse Dirichlet prior.

While BART is typically efficient in computation, processing larger datasets can lead to longer estimation times. To reduce these times, the BART program provides straightforward and accessible multi-threading options. A notable feature of efficiency in the BART package is its capability to retain trees from a previous BART model fitting, allowing for subsequent predictions using the R predict function without the need for re-fitting the model.

The BART package accommodates various types of outcomes, including continuous, binary outcomes through probit or logit transformations, categorical, and time-to-event outcomes, which cover right censoring, ab-sorbing events, competing risks, and recurrent events.

18.6.7 PRINCIPAL COMPONENT ANALYSIS

Principal Component Analysis (PCA) serves as a method for dimensionality reduction and feature selection. It is applicable in diverse fields such as pattern recognition, cluster analysis, feature extraction and selection, and the visualization of high-dimensional data. This technique ensures minimal loss of information from the input variables by retaining the most significant features, known as principal components, while reducing the dataset size and maintaining the integrity of the original data as much as possible.

The benefits of using PCA include its ability to eliminate correlated features, which are often time-consuming, contain missing variables, or are redundant due to multicollinearity, especially in large datasets. By discarding less important features, PCA not only enhances model training efficiency but also helps in mitigating overfitting by simplifying the dataset.

Currently, the critical question is determining the most suitable algorithm. Various approaches assist in deciding whether to adopt a validated model. In our study, we employ accuracy, recall, precision, f-measure, and ROC as the optimal performance metrics for classification algorithms. These metrics are particularly effective for imbalanced data, with further discussions on performance metrics to be addressed in Section 3.

18.7 PERFORMANCE EVALUATION

For the purpose of classification, the class labels are binary, specifically 'clean' and 'unclean', and the data set is labeled according to WHO standards. Many classification algorithms assess accuracy by the proportion of correctly classified observations. However, in cases of imbalanced datasets, this metric can be misleading as the minority classes have a lesser impact on the overall accuracy.

A confusion matrix encapsulates all possible outcomes of classification through TP (True Positives), TN (True Negatives), FP (False Positives), and FN (False Negatives), thereby facilitating the assessment of classification quality as shown in Figure 18.10. This matrix records the correct and incorrect classifications for each category (Shafi et al., 2018).

		Predicted	
		Positive	Negative
Actual	Positive	True Positive(TP)	False Negative(FP)
	Negative	False Positive(FP)	True Negative(TN)

FIGURE 18.10 The confusion matrix of the water qualification problem.

18.7.1 ACCURACY

Accuracy represents the proportion of correct predictions made by the model out of all observations. It is calculated using the following equation, where TP denotes true positives, TN denotes true negatives, FP denotes false positives, and FN denotes false negatives (Sokolova et al., 2006).

$$Accuracy = (TP + TN) / (TP + TN + FP + FN) \qquad (18.9)$$

Here, TP represents the count of positive samples accurately identified as positive, while TN represents the count of negative samples accurately identified as negative. In contrast, FP denotes the count of negative samples incorrectly identified as positive, and FN denotes the count of positive samples incorrectly identified as negative.

18.7.2 PRECISION

It is a measurement of correctness achieved in positive prediction that means which observations are labeled as positive and how many are actually labeled positive. The below formula is used to calculate the precision:

$$Precision = TP / (TP + FP) \qquad (18.10)$$

18.7.3 RECALL

Recall quantifies the proportion of actual positives that are correctly identified or the number of observations that are classified as positive. This metric is also known as 'Sensitivity.' The equation for computing recall is presented below:

$$Recall = TP / (TP + FN) \qquad (18.11)$$

18.7.4 F-MEASURE

F-measure is used to evaluate the effectiveness of classification by merging both precision and recall. The coefficient is calculated based on a weighted ratio that emphasizes either precision or recall. The following formula computes the F measure, where β is typically set to 1.

$$Fmeasure = \left(Recall \times precision \times (1 + \beta)^2\right) / (precision + \beta^2 \times Recall) \quad (18.12)$$

Although the methods described above surpass accuracy, they still fall short in addressing key classification issues. Precision fails to account for the accuracy of negative predictions, while Recall primarily focuses on identifying true positives.

Therefore, this analysis indicates the necessity for a more effective metric to fulfill our accuracy requirements.

18.8   RECEIVER OPERATING CHARACTERISTICS (ROC)

The ROC is a predominant metric for evaluating the effectiveness of classification models. It assesses the accuracy of predictions by plotting the True Positive (TP) rate, also known as Sensitivity, against the False Positive (FP) rate, known as Specificity. The ROC is calculated using the formula:

$$ROC = TN / (TN + FP) \qquad (18.13)$$

The ROC curve is beneficial as it visually depicts the trade-offs between the benefits (TP) and the costs (FP) in classification tasks. A larger area under the ROC curve indicates greater accuracy.

Subsequent sections will explore the experimental outcomes and their analysis for each model.

18.9   EXPERIMENTAL RESULTS AND DISCUSSION

18.9.1   PERFORMANCE ON OVERSAMPLING

The performance of models such as RF, XGBoost, DT, BART, MARS and LR has been measured on smote sampling techniques. In order to calculate the performance metrics like accuracy, f-measure, auc, and time in seconds classifier, the results after implementing the models are mentioned in Table 18.2. Compared to all the models based on the results, the RF model performed better than other classifiers, with an accuracy of 99.5%

18.9.2   PERFORMANCE ON UNDERSAMPLE

Table 18.3 presents the performance outcomes of models using different sampling techniques to address data imbalance. It shows the effectiveness of the undersampling method within the down-sampling strategy. Various machine learning models were tested using this approach to determine the most effective model for predicting water quality. The findings indicate that LR models outperformed others, achieving an accuracy of 88.2%. A comprehensive discussion of these sampling methods is provided in Section 4 under-sampling method approaches.

18.9.3   PERFORMANCE ON SMOTE

Table 18.4 displays the effectiveness of various machine learning models using the SMOTE oversampling method. It shows the accuracy rates as follows: RF at 98.4%, XGBoost at 96.4%, SVM with RBF kernel at 97.7%, DT at 92.2%, BART at 97.2%, and LR at 92%. The analysis indicates that the RF model outperformed others when

TABLE 18.2
Performance of the Models on Upsampled Technique

Performance on oversampling

Model	Accuracy	Precision	Recall	mcc	f-meas	auc-roc
Random Forest	0.995	0.991	1	0.991	0.995	1
XGboost	0.962	0.897	1	0.92	0.946	0.977
SVM RBF kernel	0.978	0.958	1	0.957	0.979	0.992
Decision Tree	0.962	0.899	1	0.92	0.946	0.977
BART	0.976	0.954	1	0.953	0.977	0.994
Logistic Regression	0.908	0.885	0.939	0.818	0.911	0.961

TABLE 18.3
Performance of the Models on Undersampled Technique

Performance on undersampled dataset

Model	Accuracy	Precision	Recall	mcc	f-meas	auc-roc
Random Forest	0.87	0.843	0.91	0.742	0.875	0.953
XGboost	0.5	0.5	1	NA	0.667	1
SVM with RBF	0.855	0.859	0.85	0.71	0.854	0.961
Decision Tree	0.712	0.718	0.7	0.425	0.709	0.742
BART	0.875	0.883	0.865	0.75	0.874	0.961
Logistic Regression	0.882	0.81	0.85	0.74	0.868	0.973

TABLE 18.4
Performance of the Models on SMOTE Technique

Performance on SMOTE

Model	Accuracy	Precision	Recall	mcc	f-meas	auc-roc
Random Forest	0.984	0.972	0.996	0.967	0.984	0.998
XGboost	0.964	0.945	0.986	0.93	0.965	0.987
SVM with RBF	0.977	0.957	1	0.956	0.978	0.993
Decision Tree	0.922	0.896	0.955	0.846	0.924	0.966
BART	0.972	0.949	0.998	0.946	0.973	0.991
Logistic Regression	0.92	0.896	0.95	0.841	0.922	0.967

using SMOTE, and details on the creation of synthetic samples are explored in the data sampling techniques section.

18.9.4 PERFORMANCE OF ROSE

ROSE oversampling was utilized to equalize the dataset and evaluate the effectiveness of various models, including RF, XGBoost, SVMRBF kernel, DT, BART, and LR. It also aimed to assess performance metrics such as accuracy, precision, recall, mcc, f-measure, and auc-roc through the use of confusion matrices for each classifier as referenced in Table 18.5. According to the data, BART and RF demonstrated superior performance compared to other methods, with a marginal lead of approximately 2%.

18.9.5 PERFORMANCE OF BSMOTE

BSMOTE, a technique belonging to the SMOTE family, is utilized for oversampling. It has been implemented in various models, including RF, XGBoost, SVMRBF kernel, DT, BART, and LR, to assess performance metrics such as accuracy, precision, recall, MCC, f-measure, and auc-roc. These metrics were derived using confusion matrices for each classifier, as detailed in Table 18.1. Based on the data presented in Table 18.5, RF has outperformed the other classifiers.

The efficacy of models, including RF, XGboost, SVM RBF kernel, DT, BART, and LR, was evaluated using the ADASYN method as shown in Table 18.7. Performance metrics such as accuracy, precision, recall, mcc, f-measure, and auc-roc were derived from confusion matrices for each classifier, as detailed in Table 18.6. The data indicates that RF outperformed the other models, achieving an accuracy of 98.1%.

TABLE 18.5
Performance of the Models on ROSE Technique

Performance on ROSE

Model	Accuracy	Precision	Recall	mcc	f-meas	auc-roc
Random Forest	0.951	0.919	0.987	0.904	0.952	0.99
XGboost	0.923	0.905	0.944	0.848	0.924	0.978
SVM RBFkernel	0.919	0.893	0.949	0.84	0.92	0.973
Decision Tree	0.821	0.652	0.997	0.688	0.788	0.881
BART	0.953	0.929	0.98	0.907	0.954	0.991
Logistic Regression	0.848	0.838	0.858	0.697	0.848	0.924

TABLE 18.6
Performance of the Models on BSMOTE Technique

Performance on b-SMOTE

Model	Accuracy	Precision	Recall	mcc	f-meas	auc-roc
Random Forest	0.982	0.983	0.98	0.963	0.982	0.996
XGboost	0.971	0.968	0.974	0.942	0.971	0.991
SVM RBFkernel	0.987	0.995	0.979	0.974	0.987	0.997
Decision Tree	0.942	0.896	0.936	0.872	0.916	0.956
BART	0.979	0.977	0.98	0.957	0.979	0.996
Logistic Regression	0.973	0.962	0.984	0.946	0.973	0.992

From Table 18.1, Table 18.3 and Table 18.6, analysis of the results concludes that on Upsample,SMOTE and ADASYN RF model performed better. The findings revealed that RF model offered the most accurate classifier with a percentage of 99.5% on Up sampling.

TABLE 18.7
Performance of the Models on ADASYN Technique

Performance of ADASYN

Model	Accuracy	Precision	Recall	mcc	f- meas	auc- roc
Random Forest	0.981	0.969	0.994	0.962	0.981	0.998
XGboost	0.5	0.5	1	NA	0.667	1
SVM RBFkernel	0.855	0.859	0.85	0.71	0.854	0.961
Decision Tree	0.712	0.718	0.7	0.425	0.709	0.742
BART	0.875	0.883	0.865	0.75	0.874	0.961
Logistic Regression	0.87	0.81	0.85	0.74	0.868	0.973

For further improvement of the model performance, we have conducted the Bayesian optimization technique and the PCA on the RF is discussed in the following section.

18.10 BAYESIAN OPTIMIZATION WITH ITERATIVE LEARNING

The objective of Bayesian optimization in problem-solving is to pinpoint the global minimum with minimal iterations. This method sequentially employs a predictive model to suggest new parameters for evaluation. In assessing potential parameter values, it forecasts the mean and variance of the results. A coefficient applied to the standard error, referred to as k, facilitates the balance between exploration and exploitation. Promising minimum locations are identified using a method called a surrogate function. An 'acquisition function' updates the prior for the surrogate function, depicted as a probability distribution. This function, which balances exploration and exploitation, prompts the generation of new testing points.

Algorithm 3 Bayesian Optimization with Iterative Learning (BOIL)

1: Input: Initial data f, Iterations N, A black box function F (x)
2: Initialize a Gaussian Process surrogate function prior distribution.
3: for each iteration i from 1 to N do
4: Choose a set of data points x such that the acquisition function a(x), utilizing the current prior distribution as input, is optimized.

5: Process the data points x to generate the results y utilizing the target cost function c(x).
6: Incorporate the new data into the Gaussian Process prior distribution to generate a posterior, which will serve as the prior for the subsequent iteration.
7: end for
8: Interpret the current Gaussian Process distribution to find the global minima.

The preceding algorithm describes the method for implementing BOIL to identify the global minimum using a surrogate function.

Table 18.2 shows the accuracy achieved of 99. 5% using the RF with Up Sampling Technique. To enhance the system's performance, we applied the BOIL technique to the RF.

Algorithm 4 Stack Ensemble Algorithm

1: Input: Dataset $D_s = \{(X_i, y_i)\}$, where i = 1 to m
2: Split the dataset D_s into training set T_r and test set T_t.
3: for each base model M_j in the ensemble do
4:    Train the model M_j on the training set T_r.
5:    Generate predictions P_j on the test set T_t.
6: end for
7: Stack the predictions of all base models to create a new dataset D_{meta}.
8: Train a meta-model on D_{meta} to make final predictions.
9: Output: Final predictions from the meta-model.

Figure 18.11 demonstrates the iterations involving PCA and upsampling, where the ROC values show significant variation from 0 to 25. Beyond this range, the results stabilize across the plot.

Figure 18.12 depicts the variations in hyperparameters across iterations, with 50 iterations set to track these changes on a scatter plot. Table 18.7 displays the hyperparameter ranges. The maximum ROC value for randomly chosen predictors occurs at 9, with the peak ROC value recorded at 20,000 trees and a minimum node size of 40.

We observed that the performance of the model did not improve with the RF with BOIL as we got the roc with 98.7%. However, we implemented the dimensionality reduction technique PCA on RF to improve the performance, which is discussed in the following section. The hyper parameters that change over the iterations are shown in Table 18.8.

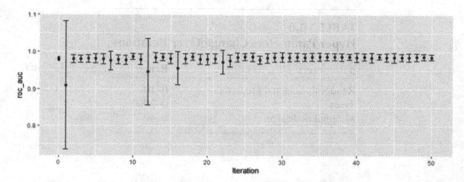

FIGURE 18.11 Plot of search iterations with RF+PCA+BOIL+Upsample.

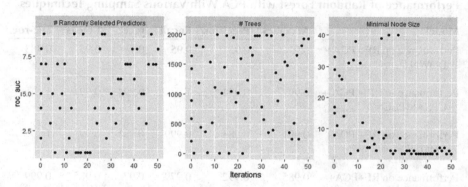

FIGURE 18.12 Plot depicting the parameters change over iterations with RF+PCA+BOIL+ Upsample.

Table 18.9 shows performance improvement in comparison with the RF with upsampling. In Table 18.1, we discussed. PCA with RF achieved better performance in the upsampling method with an accuracy of 99.7%.

Figure 18.13 illustrates the performance of the models on the various sampling techniques on RF with PCA. We notice that the performance of RF using PCA on SMOTE data and ADASYN data are similar. However, there is a marginal increase in the performance of the model when executed on features extracted using PCA on upsampled data.

18.11 CONCLUSION

In this work, we identified the problem of classification of water quality and created ML approaches on a binary classification dataset. So, as the data set was imbalanced, various types of sampling strategies such as upsampling, downsampling, SMOTE, ROSE, ADASYN, and b-SMOTE were implemented on the data set. After getting the required results, the models were subjected to dimensionality reduction using PCA,

TABLE 18.8
Hyper Parameters Change Over Iterations

parameters	Range
Randomly selected predictors	0–10
Trees	0–2000
Minimal node size	0–40

TABLE 18.9
Performance of Random Forest with PCA With Various Sampling Techniques

Model	Accuracy	Precision	Recall	mcc	f-meas	auc-roc
Performance on RF+ PCA+ DOWN	0.95	0.95	0.95	0.9	0.95	0.983
Performance in RF + PCA + UPSAMPLE	0.997	0.994	1	0.994	0.997	1
Performance on RF+ PCA+ SMOTE	0.991	0.983	0.999	0.981	0.991	1
Performance on RF+PCA+ b-SMOTE	0.985	0.992	0.979	0.97	0.985	0.999
Performance in RF + PCA + ROSE	0.966	0.95	0.985	0.933	0.967	0.993
Performance on RF+ PCA+ ADASYN	0.99	0.982	0.999	0.981	0.99	1

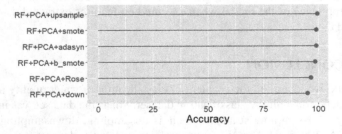

FIGURE 18.13 Accuracy comparison on RF+ PCA+ sampling methods.

and we also tried to improve the performance using the iterative Bayesian optimization approach. After investigating all the accuracies and the performance of the models on different parameters, we identified the performance in the RF as superior on the upsampling data with different splitting criteria of 90% and 10%, which performed best with 99.7%. We also show the results of AUC in different iterations using iterative Bayesian optimization. In this work, we have limited the work to the binary classification problem. But in the future, we intend to create adaptive ML approaches that are deployed online on data streams to predict water quality in real-time.

ACKNOWLEDGMENTS

The authors thank the Center of Excellence in Internet of Things, VITAP University, for its support in the completion of this work.

NOTE

1   https://kaggle.com/anbarivan/indian-water-quality-data

REFERENCES

Abuzir, S. Y., & Abuzir, Y. S. (2022). Machine learning for water quality classification. *Water Quality Research Journal*, 57(3), 152–164.

Abyaneh, H. Z. (2014). Evaluation of multivariate linear regression and artificial neural networks in prediction of water quality parameters. *Journal of Environmental Health Science and Engineering*, 12(1), 1–8. https://doi.org/ 10.1186/2052-336x-12-40

Advantages of XGBoost Algorithm in Machine Learning. http://theprofessionalspoint.blogs pot.com/ 2019/03/advantages-of-xgboost-algorithm-in.html.

Arun, J. V., & Premkumar, A. (2021). Health impacts of contaminated water in India: Coping strategies for sustainable development. In *Strategies and Tools for Pollutant Mitigation: Avenues to a Cleaner Environment*, 391–403.

Asadollah, S. B. H. S., Sharafati, A., Motta, D., & Yaseen, Z. M. (2021). River water quality index prediction and uncertainty analysis: A comparative study of machine learning models. *Journal of Environmental Chemical Engineering*, 9(1), 104599. https://doi.org/ 10. 1016/j.jece.2020.104599

Asselman, A., Khaldi, M., & Aammou, S. (2023). Enhancing the prediction of student performance based on the machine learning XGBoost algorithm. *Interactive Learning Environments*, 31(6), 3360–3379.

Azrour, M., Mabrouki, J., Fattah, G., Guezzaz, A., & Aziz, F. (2022). Machine learning algorithms for efficient water quality prediction. *Modeling Earth Systems and Environment*, 8(2), 2793–2801. https://doi.org/10.1007/s40808-021-01266-6

Bunkhumpornpat, C., Sinapiromsaran, K., & Lursinsap, C. (2009). Safe-Level-SMOTE: Safe-level-Synthetic Minority Over-Sampling Technique for handling the class imbalanced problem. In *Advances in Knowledge Discovery and Data Mining*, pp. 475–482. Springer Berlin Heidelberg. https://doi.org/10.1007/978-3-642-01307-2_43

Chawla, N. V., Bowyer, K. W., Hall, L. O., & Kegelmeyer, W. P. (2002). SMOTE: Synthetic minority over-sampling technique. *Journal of Artificial Intelligence Research*, 16, 321–357.

Chen, S., Hong, X., & Harris, C. J. (2010). Particle swarm optimization aided orthogonal forward regression for unified data modeling. *IEEE Transactions on Evolutionary Computation*, 14(4), 477–499.

Chen, T., & Guestrin, C. (2016, August). Xgboost: A scalable tree boosting system. In *Proceedings of the 22nd ACM SIGKDD International Conference on Knowledge Discovery and Data Mining*, pp. 785–794.

Das Kangabam, R., Bhoominathan, S. D., Kanagaraj, S., & Govindaraju, M. (2017). Development of a water quality index (WQI) for the Loktak Lake in India. *Applied Water Science*, 7(6), 2907–2918. www.kaggle.com/anbarivan/indian-water-quality-data. /datasets/anbarivan/indian-water-quality-data

Donders, A. R. T., Van Der Heijden, G. J., Stijnen, T., & Moons, K. G. (2006). A gentle introduction to imputation of missing values. *Journal of Clinical Epidemiology*, 59(10), 1087–1091.

Elreedy, D., & Atiya, A. F. (2019). A comprehensive analysis of Synthetic Minority Oversampling Technique (SMOTE) for handling class imbalance. *Information Sciences*, 505, 32–64. https://doi.org/10.1016/j.ins.2019.07.070

Gao, M., Hong, X., Chen, S., Harris, C. J., & Khalaf, E. (2014). PDFOS: PDF estimation based over-sampling for imbalanced two-class problems. *Neurocomputing*, 138, 248–259. https://doi.org/10.1016/j.neucom.2014.02.006

Goparaju, S. U. N., Vaddhiparthy, S. S. S., Pradeep, C., Vattem, A., & Gangadharan, D. (2021, June). Design of an IoT system for machine learning calibrated TDS measurement in smart campus. In *2021 IEEE 7th World Forum on Internet of Things (WF-IoT)*. https:// doi.org/10.1109/wf-iot51360.2021.9595057

Haghiabi, A. H., Nasrolahi, A. H., & Parsaie, A. (2018). Water quality prediction using machine learning methods. *Water Quality Research Journal*, 53(1), 3–13.

Han, H., Wang, W. Y., & Mao, B. H. (2005, August). Borderline-SMOTE: A new over-sampling method in imbalanced data sets learning. In *Lecture Notes in Computer Science* (pp. 878–887). Springer Berlin Heidelberg. https://doi.org/10.1007/ 11538059_91

Han, J., Pei, J., & Tong, H. (2022). *Data Mining: Concepts and Techniques*. Morgan kaufmann.

Harikumar, P. S., Aravind, A., & Vasudevan, S. (2017). Assessment of water quality status of Guruvayur municipality. *Journal of Environmental Protection*, 8(2), 159–170.

He, H., Bai, Y., Garcia, E. A., & Li, S. (2008, June). ADASYN: Adaptive synthetic sampling approach for imbalanced learning. In *2008 IEEE International Joint Conference on Neural Networks (IEEE World Congress on Computational Intelligence)*, pp. 1322–1328. IEEE. https://doi.org/10.1109/ijcnn.2008.4633969

Henderi, H., Wahyuningsih, T., & Rahwanto, E. (2021). Comparison of Min-Max normalization and Z-score normalization in the K-nearest neighbor (kNN) algorithm to test the accuracy of types of breast cancer. *International Journal of Informatics and Information Systems*, 4(1), 13–20.

Jaloree, S., Rajput, A., & Gour, S. (2014). Decision tree approach to build a model for water quality. *Binary Journal of Data Mining & Networking*, 4(1), 25–28.

Jung, H., Senf, C., Jordan, P., & Krueger, T. (2020). Benchmarking inference methods for water quality monitoring and status classification. *Environmental Monitoring and Assessment*, 192(4), 1–17.

Kedia, N. (2015, September). Water quality monitoring for rural areas – A Sensor Cloud based economical project. In *2015 1st International Conference on Next Generation Computing Technologies (NGCT)*, pp. 50–54. IEEE. https://doi.org/10. 1109/ngct.2015.7375081

Khan, M. S. I., Islam, N., Uddin, J., Islam, S., & Nasir, M. K. (2022). Water quality prediction and classification based on principal component regression and gradient boosting classifier approach. *Journal of King Saud University–Computer and Information Sciences*, 34(8), 4773–4781.

Khullar, S., & Singh, N. (2021). Machine learning techniques in river water quality modelling: a research travelogue. *Water Supply*, 21(1), 1–13. https://doi.org/10.2166/ws.2020.277

Koditala, N. K., & Pandey, P. S. (2018, August). Water quality monitoring system using IoT and machine learning. In *2018 International Conference on Research in Intelligent and Computing in Engineering (RICE)*, pp. 1–5. IEEE.

Krhoda, G., & Amimo, M. O. (2019). Groundwater quality prediction using logistic regression model for Garissa County. *Africa Journal of Physical Sciences*, 3, 13–27.

Ladjal, M., Ouali, M. A., & Lass, M. D. (2020, September). Optimization of SVM parameters with hybrid PCA-PSO methods for water quality monitoring. In *2020 International Conference on Electrical Engineering (ICEE)*, pp. 1–6. IEEE. https://doi.org/10.1109/icee49691.2020.9249881

Li, D. C., Wang, S. Y., Huang, K. C., & Tsai, T. I. (2022). Learning class-imbalanced data with region-impurity Synthetic Minority Oversampling Technique. *Information Sciences*, 607, 1391–1407.

Lunardon, N., Menardi, G, & Torelli, N. (2014). ROSE: A package for binary imbalanced learning (2014). *R Journal*, 6(1). https://cran.r-project.org/web/packages/smotefamily/smotefamily.pdf

Madley-Dowd, P., Hughes, R., Tilling, K., & Heron, J. (2019). The proportion of missing data should not be used to guide decisions on multiple imputation. *Journal of Clinical Epidemiology*, 110, 63–73.

Muharemi, F., Logofătu, D., & Leon, F. (2019). Machine learning approaches for anomaly detection of water quality on a real-world data set. *Journal of Information and Telecommunication*, 3(3), 294–307. https://doi.org/10.1080/24751839.2019. 1565653

Mustaffa, Z., & Yuhanis, Y. (2010). A comparison of normalization techniques in predicting dengue outbreak. *International Conference on Business and Economics Research*, 1, 345–349.

Nasir, N., Kansal, A., Alshaltone, O., Barneih, F., Sameer, M., Shanableh, A., & Al-Shamma'a, A. (2022). Water quality classification using machine learning algorithms. *Journal of Water Process Engineering*, 48, 102920.

Radhakrishnan, N., & Pillai, A. S. (2020, June). Comparison of water quality classification models using machine learning. In *2020 5th International Conference on Communication and Electronics Systems (ICCES)*, pp. 1183–1188. IEEE. https://doi.org/10.1109/icces48766.2020.9137903

Raju, V. G., Lakshmi, K. P., Jain, V. M., Kalidindi, A., & Padma, V. (2020, August). Study the influence of normalization/transformation process on the accuracy of supervised classification. In *2020 Third International Conference on Smart Systems and Inventive Technology (ICSSIT)*, pp. 729–735. IEEE. https://doi.org/10.1109/icssit48917.2020.9214160

Rasaei, Z., & Bogaert, P. (2019). Spatial filtering and Bayesian data fusion for mapping soil properties: A case study combining legacy and remotely sensed data in Iran. *Geoderma*, 344, 50–62. https://doi.org/10.1016/j.geoderma. 2019.02.031

Sahour, S., Khanbeyki, M., Gholami, V., Sahour, H., Kahvazade, I., & Karimi, H. (2023). Evaluation of machine learning algorithms for groundwater quality modeling. *Environmental Science and Pollution Research*, 30(16), 46004–46021.

Shafi, U., Mumtaz, R., Anwar, H., Qamar, A. M., & Khurshid, H. (2018, October). Surface water pollution detection using Internet of Things. In *2018 15th International Conference on Smart Cities: Improving Quality of Life Using ICT & IoT (HONET-ICT)*, pp. 92–96. IEEE.

Shaibur, M. R., Hossain, M. S., Khatun, S., & Tanzia, F. S. (2021). Assessment of drinking water contamination in food stalls of Jashore Municipality, Bangladesh. *Applied Water Science*, 11(8), 142.

Shakhari, S., & Banerjee, I. (2019). A multi-class classification system for continuous water quality monitoring. *Heliyon*, 5(5), e01822.

Singh, D., & Singh, B. (2020). Investigating the impact of data normalization on classification performance. *Applied Soft Computing*, 97, 105524.

Sokolova, M., Japkowicz, N., & Szpakowicz, S. (2006, December). Beyond accuracy, F-score and ROC: a family of discriminant measures for performance evaluation. In *Australasian Joint Conference on Artificial Intelligence*, pp. 1015–1021. Berlin, Heidelberg: Springer Berlin Heidelberg.

Sparapani, R., Spanbauer, C., & McCulloch, R. (2021). Nonparametric machine learning and efficient computation with Bayesian Additive Regression Trees: The BART Package. *Journal of Statistical Software*, 97, 1–66. https://doi.org/10.18637/jss.v097.i01

Tu, J. V. (1996). Advantages and disadvantages of using artificial neural networks versus logistic regression for predicting medical outcomes. *Journal of Clinical Epidemiology*, 49(11), 1225–1231. https://doi.org/10. 1016/s0895-4356(96)00002-9

Tyagi, S., & Mittal, S. (2020). Sampling approaches for imbalanced data classification problem in machine learning. In *Proceedings of ICRIC 2019: Recent innovations in computing*, pp. 209–221. Springer International Publishing. https://doi.org/10. 1007/978-3-030-29407-6_17

Vasudevan, S. K., & Baskaran, B. (2021). An improved real-time water quality monitoring embedded system with IoT on unmanned surface vehicle. *Ecological Informatics*, 65, 101421.

Vida, A. (2016). Practical guide to deal with imbalanced classification problems in R. imbalanceAnalyticsVida. www.analyticsvidhya.com/back-channel/download-pdf.php?pid=24221

Wang, F., Wang, Y., Zhang, K., Hu, M., Weng, Q., & Zhang, H. (2021). Spatial heterogeneity modeling of water quality based on random forest regression and model interpretation. *Environmental Research*, 202, 111660. https://doi.org/10.1016/j.envres.2021.111660

Wu, W., May, R., Dandy, G. C., & Maier, H. R. (2012). A method for comparing data splitting approaches for developing hydrological ANN models.

Yu, W. M., Du, T., & Lim, K. B. (2004, December). Comparison of the support vector machine and relevant vector machine in regression and classification problems. In *ICARCV 2004 8th Control, Automation, Robotics and Vision Conference*, vol. 2, pp. 1309–1314. IEEE.

Zahin, S. A., Ahmed, C. F., & Alam, T. (2018). An effective method for classification with missing values. *Applied Intelligence*, 48(10), 3209–3230.

Zamri, N., Pairan, M. A., Azman, W. N. A. W., Abas, S. S., Abdullah, L., Naim, S., ... & Gao, M. (2022). River quality classification using different distances in k-nearest neighbors algorithm. *Procedia Computer Science*, 204, 180–186. https://doi.org/10.1016/j.procs.2022.08.022

19 Ethical Considerations and Social Implications

Jeethu V. Devasia, Deepanramkumar P.,
Helensharmila A., and Gokul Yenduri

19.1 INTRODUCTION

Agriculture has a rich history that spans thousands of years. The shift from a nomadic, hunter-gatherer lifestyle to settled agriculture marks a crucial turning point in human civilization. The ability to cultivate plants and domesticate animals allowed communities to establish permanent settlements, leading to the development of complex societies and civilizations. Agriculture plays a pivotal role in ensuring food security by providing a stable and consistent supply of food. The cultivation of crops and the domestication of animals allow for a reliable source of sustenance. Different regions and cultures have developed unique agricultural techniques based on climate, soil conditions, and available resources. From the rice paddies of Asia to the wheat fields of the Middle East, the diversity in agricultural practices reflects the adaptability of humans to various environments. Over the centuries, agriculture has witnessed numerous technological advancements. From the plow and irrigation systems to modern machinery and precision farming technologies, innovation has continually improved efficiency and productivity [1].

Recent advancements in three crucial domains have significantly expanded the potential of data-driven farming. These include progress in data generation, such as mobile devices, field sensors, satellites, and the engagement of farmers as sensors; advancements in data processing and predictive analytics, employing big data stacks, machine learning (ML), and deep learning; and improvements in human-computer interactions. The latter involves human-centric approaches aimed at creating experiences that enhance the ease and utilization of insights through voice, text, and images [2].

In the early stages of smart farming, diverse sets of data related to various aspects of agriculture, such as weather conditions, soil quality, animal behavior, and harvest monitoring, are gathered. The main strategies that can be used for data collection are sensor networks and satellite imagery.

- Sensor networks: Farms use various sensors to collect data on soil moisture, temperature, crop health, etc. The widespread use of these sensors raises concerns about the privacy of farmers and their operations.

DOI: 10.1201/9781003484608-19

- Satellite imagery: Remote sensing technologies, including satellite imagery, are used to monitor large agricultural areas. The analysis of such imagery can reveal sensitive information about farming practices.

Advancements in technologies and adaptation of technologies such as ML in agriculture aid in the development of large-scale farming, contributing to a sustainable rural economy [3]. It also provides smart solutions for disease recognition in crops like cotton, promoting sustainable agriculture [4]. Additionally, ML enables better crop prediction based on regional weather data, thus supporting informed agricultural decisions [5]. A brief discussion on various technological aspects of farming follows next.

Personalizing Sustainable Agriculture with Causal ML: This approach employs advanced algorithms to analyze environmental data and predict the best agricultural practices for specific conditions. It can also factor in economic considerations, like cost-benefit analyses of different farming techniques, leading to more profitable and sustainable practices. Furthermore, this technology facilitates a deeper understanding of the long-term impact of agricultural decisions on soil health and ecosystem balance, ensuring sustainable farming not just in the present but for future generations as well [6].

Computer Vision and ML for Smart Farming and Agriculture Practices: By integrating these technologies, farmers can automate and optimize processes such as planting, harvesting, and sorting. The system can analyze crop health at different growth stages, providing insights for timely intervention. This technology is also pivotal in large-scale monitoring of agricultural lands, enabling precise and efficient management of vast areas with minimal human intervention, thus reducing labor costs and increasing overall efficiency [7].

ML Applications in Agriculture: The range of ap-plications also includes advanced forecasting models for weather and market demands, which are crucial for strategic planning in agriculture. These models help in predicting the best times to plant and harvest and in determining the most profitable crops to grow. Furthermore, ML algorithms can assist in the genetic engineering of crops, making them more resistant to diseases and adaptable to changing climatic conditions [8].

Machine Learning-Based Agriculture: Predictive analytics in agriculture also extends to livestock management, where it can be used for monitoring the health and productivity of animals. This involves analyzing data from various sensors to detect early signs of illness, thereby preventing outbreaks. Additionally, ML models can optimize feed formulas to enhance the growth and health of livestock, contributing to more sustainable animal husbandry practices [9].

Sustainable Agriculture through Data Analytics: The use of data analytics also extends to tracking and reducing the carbon footprint of farming activities. By analyzing energy consumption patterns and resource utilization, these tools can suggest more environmentally friendly practices. They also play a crucial role in sustainable water management by predicting irrigation needs and optimizing water usage, thus conserving this vital resource [10].

The integration of technology into farming practices enhances the overall agricultural ecosystem. Smart devices, sensors, and automated machinery contribute to a more connected and efficient farming environment. Training systems using ML algorithms allow for continuous improvement and adaptation to changing conditions [11].

Despite the benefits, there are challenges, such as high initial costs, concerns about data privacy, and the need for farmer education. Addressing these challenges is crucial for the widespread adoption of machine learning-based technologies in agriculture. Certainly, discussing the ethical considerations and social implications of using ML technologies in farming is important, as these technologies continue to play an increasingly prominent role in agriculture.

In this chapter, we delve into the ethical aspects and societal consequences associated with the utilization of ML technologies in agriculture. The focus is on exploring the responsible deployment of these technologies and understanding their impact on both ethical considerations and social dynamics within the farming sector.

19.2 ETHICAL CONSIDERATIONS

Addressing ethical considerations in sustainable farming through ML involves careful examination of various aspects to ensure responsible and beneficial implementation. Some key ethical considerations in the context of sustainable farming with ML are discussed next.

19.2.1 DATA PRIVACY

Smart devices and technologies play a crucial role in enabling farmers to gather and analyze data pertaining to various facets of their operations. This data provides valuable insights for multiple stakeholders, including farmers, technology providers, supply chain analysts, and agricultural service providers. Given the substantial volume, rapid pace, and diverse nature of these data sources, they can be categorized as big data. The extensive adoption of technologies for data collection and communication has raised privacy apprehensions regarding farmers and the confidentiality of their data [12]. We have to ensure that the data collected through ML applications in farming is handled with the utmost respect for privacy rights. Farmers' personal and operational data should be anonymized and secured to prevent unauthorized access.

Other major concerns are with respect to data ownership and control.

- Farmer's rights: The ownership and control of agricultural data, especially when collected by third-party providers or ML platforms, pose challenges. Farmers need assurance that their data will not be used without their consent.
- Data sharing: Collaborative efforts and data sharing between farms for better ML models might lead to concerns about the privacy and security of shared data.

There are numerous techniques that may contribute to ensuring data privacy in machine learning-based applications. Some of the most popular privacy-preserving strategies are based on cryptography and statistics. A brief description of the main approaches is given next.

k-anonymity, l-diversity, and t-closeness are different models for protecting privacy by adding some anonymization to the data before disclosure [13], [14], [15]. Multi-party computation-based federated learning allows a number of parties to collect-ively train models without revealing private datasets [16]. The majority of federated learning systems employ different methods, such as differential privacy [17], [18], secure multi-party computation [19], [20], homomorphic encryption [21] etc., as add-itional measures. A detailed discussion on privacy-preserving solutions in the big data lifecycle and privacy-preserving technologies in smart farming is given in [12]. The paper gives an overview of multiple technologies that can be used in the farming sector to address privacy concerns, including a combination of principal component analysis and Bayesian estimation.

19.2.2 TRANSPARENCY AND ACCOUNTABILITY

Transparency and accountability can be considered as another major segment under ethical considerations. In general, transparency is a fundamental principle that involves sharing information, decisions, and processes openly with relevant stakeholders. It fosters trust and mutual understanding between different parties. Transparency goes hand-in-hand with accountability. When there is transparency in actions and decisions, there will be accountability for the outcomes [22]. In order to maintain transparency and accountability, clear information on how ML algorithms make decisions in farming processes should be provided. Transparent models allow farmers and stakeholders to understand, trust, and verify the technology's outcomes.

Integrating explainable artificial intelligence techniques is a promising approach for addressing transparency and accountability concerns [23]. Explainable artificial intelligence refers to the ability of an artificial intelligence system to provide under-standable and interpretable explanations for decisions or outputs. In terms of trans-parency, a better grasp of how the technology adopted in farming works and why it makes specific predictions or decisions shall be explained. Explainable artificial intel-ligence aims to provide explanations that are understandable to humans, even those without a technical background. This involves translating complex AI model outputs into language or visualizations that are accessible to a broader audience. Providing explanations helps ensure that AI systems are accountable and can be audited for biases or unintended consequences.

Several techniques are employed in explainable artificial intelligence, including rule-based systems, feature importance analysis, model-agnostic methods like LIME (Local Interpretable Model-agnostic Explanations), SHAP (SHapley Additive exPlanations), and various visualization tools [11].

19.2.3 ADDRESSING THE DIGITAL DIVIDE

The digital divide in agriculture refers to the gap between those who have access to digital technologies, including ML tools, and those who do not. Addressing this divide is crucial for ensuring that the benefits of ML in agriculture are accessible to all farmers, regardless of their location, resources, or technological capabilities

[2]. When digitalization and automation become mainstream, the lives of workers and farmers who can afford these technologies will be substantially improved. Technology-based data-driven farming involves utilizing data to enhance decision-making within farming systems, leading to improvements in various aspects of the food system, including crop yields, profits, environmental sustainability, and food security.

The attainment of these solutions requires essential prerequisites, namely, widespread mobile network coverage, ownership of mobile handsets, and access to affordable mobile service subscriptions. Despite recent expansions in the reach of mobile networks, increased mobile phone ownership, and reduced costs of mobile data, significant access gaps persist. It is estimated that almost half of the global population still lacks internet access. This digital divide raises concerns about potentially impeding the realization of human rights and the United Nations' Sustainable Development Goals, particularly those pertaining to education, equity, health, and well-being [24].

19.2.4 BALANCING TECHNOLOGY AND ENVIRONMENTAL CONSERVATION

Balancing technology and environmental conservation is crucial when incorporating ML in agriculture. While ML can offer innovative solutions to enhance agricultural practices, it is essential to ensure that these technologies are deployed sustainably, minimizing negative impacts on the environment.

The introduction of the conservation agriculture approach aimed to enhance and sustain productivity in agroecosystems, fostering increased profits for farmers while concurrently preserving natural resources. This approach involves the effective management of natural resources across farm, village, and landscape scales, with the goal of fostering synergies between food production and the conservation of ecosystems [25]. The cropping system based on conservation agriculture and the holistic approach to farming not only preserves natural resources but also facilitates cost-effective production, enhances soil health, supports timely planting, ensures crop diversification, and mitigates environmental pollution as well as the negative impacts of climate change.

19.3 SOCIAL IMPLICATIONS

The social implications of sustainable farming through ML encompass various aspects, such as enhancing rural economies, improving agricultural practices, and addressing challenges in crop management.

19.3.1 JOB DISPLACEMENT AND TRAINING

As ML techniques are revolutionizing the field of sustainable farming, challenges related to job displacement and training are to be addressed. The implementation of ML in agriculture facilitates the generation of high-quality training data, which is crucial for developing efficient ML-based query optimizers. These optimizers play a significant role in estimating costs and cardinalities in farming practices, enhancing the overall productivity and sustainability of the sector. The active learning approach, involving human intervention and feedback, ensures the generation of more accurate and relevant data for these systems. This intersection of human expertise and machine intelligence leads to more effective training programs and decision-making processes in agriculture [9].

The application of ML in sustainable farming goes beyond mere technological advancement; it extends to transforming the workforce and job roles in the agricultural sector. By automating routine tasks, ML allows farm workers to focus on more complex and strategic aspects of farming. This shift necessitates a retraining of the workforce, where traditional farming skills are augmented with digital literacy and ML competency. Training programs, therefore, need to evolve to equip farmers and agricultural workers with the skills necessary to operate and interact with ML-driven systems. This evolution in training not only enhances job satisfaction but also prepares the workforce for future technological advancements [26].

Furthermore, ML applications in agriculture extend to decision-making and precision farming, where data-driven insights allow for more targeted and efficient farming practices. By analyzing vast amounts of data, ML algorithms can provide recommendations on crop selection, pest control, resource allocation, and more. This level of precision not only improves yield and reduces waste but also promotes environmentally friendly practices. It enables farmers to understand their practices better and adapt to changing environmental conditions, ultimately leading to more sustainable farming methods [8].

In addition to improving agricultural practices, ML also reveals hidden features and patterns that may be overlooked by traditional methods. This capability is particularly useful in evaluating sustainable training practices and predicting key predictors for successful farming outcomes. For instance, ML can identify which training methods are most effective for different types of farms or farmers, leading to more personalized and effective training programs. This tailored approach not only improves the learning experience for farmers but also ensures that the training is directly applicable and beneficial to their specific farming practices [27].

Lastly, the integration of ML in farming is not just about technological upgrades; it is about a holistic transformation of the agricultural sector toward sustainability and efficiency. By harnessing the power of ML, sustainable farming can become more adaptive, resilient, and responsive to the challenges posed by climate change and a growing global population. The continuous development and refinement of ML algorithms, coupled with ongoing training and skill development, will be key to achieving these goals. As the sector evolves, so too must the tools and techniques used to train and support the agricultural workforce, ensuring that they are well-equipped to meet the demands of modern, sustainable farming [28].

19.3.2 ENSURING INCLUSIVE TECHNOLOGY ADOPTION

The concept of "Ensuring Inclusive Technology Adoption for Sustainable Farming through ML" encompasses various facets of modern agriculture. This approach aims to leverage the power of ML and related technologies to transform traditional farming practices into more efficient, sustainable, and inclusive operations.

Innovative Crop Recommendation Systems: The integration of ML and IoT (Internet of Things) in agriculture has led to the development of advanced crop recommender systems. These systems analyze various data points, such as soil conditions, weather patterns, and crop characteristics, to provide tailored recommendations to farmers. This approach ensures that farmers, regardless of their technical expertise, can benefit from precision agriculture, leading to more sustainable farming practices. The implementation of such systems is a significant step toward inclusive technology adoption in agriculture [29].

Barriers to ML Adoption: Despite the potential benefits, there are barriers to the widespread adoption of ML-based agricultural decision support systems. Understanding farmers' engagement with these technologies is crucial. It involves addressing factors such as the digital literacy of farmers, the accessibility of technology, and the relevance of the technology to local farming practices. Overcoming these barriers is key to ensuring that the benefits of ML in agriculture are accessible to all farmers, thereby promoting inclusive and sustainable agricultural practices [30].

Smart Farming through Computer Vision and ML: The use of computer vision alongside ML in agriculture is transforming farming practices. This technology enables the monitoring of crops and livestock with greater accuracy and efficiency. It can identify pests, diseases, and nutritional deficiencies, allowing for timely and precise intervention. This level of precision in monitoring and intervention ensures that sustainable farming practices can be adopted widely, including by those with limited resources or knowledge [7].

Integrative Approaches for Sustainable Solutions: An integrative approach that combines various data sources and ML algorithms is essential for recommending sustainable farming solutions. This method involves not just technological solutions but also considers economic, environmental, and social factors. Such an approach ensures that recommendations are not only technically sound but also feasible and beneficial for diverse farming communities. It is a step toward making sustainable agriculture more inclusive and accessible to all farmers [31].

Responsible Innovation for Precision Agriculture: The pursuit of responsible innovation in precision agriculture technologies is vital. This in-volves developing technologies that are not only effective but also ethical and accessible. Responsible innovation takes into consideration the environmental impact, social implications, and economic feasibility of technology adoption in agriculture. Ensuring that precision agriculture technologies powered by ML are developed and implemented responsibly is key to achieving inclusive and sustainable farming practices [32].

19.3.3 RESILIENCE AND DEPENDENCY

The intersection of resilience and dependency in sustainable farming through ML encompasses several vital aspects of modern agriculture. This concept involves using ML to enhance the robustness of farming practices against environmental challenges while also considering the dependency on technology.

Large-Scale Farming and Cyber Risk Analysis: The development of large-scale farming supported by ML is transforming rural economies. This approach, which emphasizes explainable ML models, is crucial in safeguarding the digital backbone of modern agriculture. By focusing on cyber risk analysis, these models ensure the security and integrity of agricultural data systems. This protection is vital for maintaining the resilience of farming operations against data breaches, which is essential for the sustainable growth of technology-based farming systems [33].

Big Earth Data and ML for Resilient Agriculture: The integration of Big Earth Data with ML techniques presents a powerful strategy for sustainable and resilient agriculture. This combination allows for the detailed analysis of environmental data, equipping farmers with critical insights to adapt to climate change. The data-driven decisions facilitated by this integration lead to enhanced crop yields and efficient resource management, showcasing how technology can fortify farming systems against environmental challenges [34].

Weather Prediction for Hostile Climate Adaptation: In regions facing extreme weather conditions, like the Sahel, ML-based weather prediction models are crucial for agricultural resilience. These models provide farmers with accurate forecasts, enabling them to plan and adapt to climatic adversities. This predictive capability is fundamental for strategic agricultural planning, such as irrigation and harvesting schedules, reducing the risks associated with climate change and ensuring food security in vulnerable regions [35].

Smart Farming through Computer Vision and ML: Smart farming, which combines computer vision and ML, is revolutionizing how agricultural practices are conducted. This technology enables precise monitoring of crop health, leading to more informed and immediate decision-making. The real-time insights provided help address various agricultural challenges promptly, enhancing crop yields while reducing the environmental impact. This approach represents a significant stride toward achieving sustainable agriculture [7].

ML in Designing Sustainable Agriculture Systems: Applying ML and soft computing in the design of sustainable agriculture systems marks a significant advancement in the field. These systems aim to optimize resource use and reduce environmental impacts while maintaining high crop productivity. By analyzing diverse data sets, ML algorithms provide tailored recommendations for farming practices, aligning with the overarching goals of sustainable agriculture to preserve resources for future generations [36].

19.3.4 COMMUNITY ENGAGEMENT

Community engagement plays a crucial role in promoting sustainable farming. By involving the community in farming activities, such as urban agriculture and green marketing, positive outcomes can be achieved in terms of economic,

social, and environmental sustainability. Studies have shown that community engagement positively influences expectations for the sustainability of urban agriculture. Additionally, community engagement is essential for the successful implementation of projects related to sustainable urbanism, as it fosters resilient outcomes and addresses social challenges. To ensure the creation of a sustainable environment, community engagement should be based on a thor-ough understanding of the community's needs and can be facilitated through information dissemination, inclusion, and identification. By eliminating intermediaries and incorporating green marketing principles, direct engagement between producers and consumers can be established, leading to increased profits for farmers and improved access to farm produce for consumers [37].

19.4 CASE STUDY: ETHICAL AND SOCIETAL IMPLICATIONS OF AI AND ML

This research study examines various ethical concerns, such as bias, transparency, accountability, and privacy, which arise in the development and deployment of AI and ML technologies. It also analyzes the societal impacts, including implications for employment, economic inequality, and social cohesion. The study underscores the importance of regulation and governance to ensure the responsible development and use of these technologies, highlighting the need for ethical and societal considerations in their deployment [38].

The study begins by acknowledging the advancements brought by AI and ML in various fields while emphasizing the need to consider their ethical and societal implications. It points out concerns like potential biases in algorithms, loss of privacy and autonomy, and the risk of job displacement due to automation.

19.4.1 KEY ETHICAL AND SOCIETAL CONCERNS

- *Bias in Algorithms:* The study discusses how AI and ML systems, often trained on datasets reflecting societal biases, can reinforce and perpetuate these biases in their decisions.
- *Job Displacement:* The potential for AI and ML systems to perform tasks previously done by humans raises concerns about widespread job displacement in various industries.
- *Concentration of Power:* The control of AI and ML development and deployment by a few large tech companies leads to concerns over accountability and transparency.
- *Privacy and Autonomy:* The ability of AI and ML systems to collect and analyze vast amounts of personal data poses significant privacy risks.

19.4.2 SOLUTIONS AND RECOMMENDATIONS

The study proposes solutions such as developing fair and transparent AI systems, implementing regulations to protect privacy and autonomy, and promoting retraining and upskilling programs to mitigate job displacement. These solutions aim to address the ethical and societal challenges posed by AI and ML.

19.4.3 FINAL OBSERVATIONS

The study observes that while AI and ML offer significant benefits, their ethical and societal implications, such as bias, job displacement, privacy concerns, and concentration of power, require careful consideration and responsible management. The study suggests that addressing these concerns through fair and transparent systems, privacy protection, and job displacement mitigation is essential for the responsible use of AI and ML technologies. This detailed analysis provides a comprehensive overview of the ethical and societal implications of AI and ML, highlighting the importance of considering these aspects in the advancement of these technologies.

19.5   REGULATORY FRAMEWORKS

Different organizations have taken the initiative to formulate regulatory frameworks for the promising yet challenging adoption of ML technologies in making a smart world. Here, we provide a briefing on the initiatives taken by the United Nations and India.

The Food and Agriculture Union of the United Nations has published the Regulatory Framework for Agricultural Data in the Near East and North Africa Region under Sustainable Development Goals [39]. It contains discussion on various aspects, starting from 'Impact of big data in agriculture' and 'Digital agriculture – A continuous transformation of agriculture' to 'Recommendations on harmonizing the regulatory frameworks and systems on agricultural data in the Near East and North Africa (NENA region)' and 'Recommendations on how to use the agricultural data to support the transformation of the agrifood systems in the NENA region'. The primary focus of regulations lies in safeguarding the ownership of agricultural data and the associated rights to derived data. This can be accomplished either by enacting specific legislation dedicated to farm data or by modifying existing laws related to intellectual property, data protection, antitrust, and competition. Establishing agreements for sharing data, which outline the terms and conditions governing the sharing and utilization of data, can contribute to ensuring equitable distribution of the value generated from agricultural data among all involved stakeholders. Liability mechanisms play a crucial role in the governance of agricultural data, ensuring responsible and ethical use and addressing any harm resulting from the inappropriate handling of data. Various aspects, such as data privacy and security, data sharing and interoperability, data standardization and quality, regular evaluation, etc., are addressed. The framework also mentions that striking a proper balance between adjusting to evolving technological trends and establishing a fair, competitive environment is essential for the sustainable and just progress of the agricultural industry.

NITI Aayog (National Institution for Transforming India) has proposed a manual to establish the foundation for developing the National Strategy for Artificial Intelligence in India [40]. Along with a brief description of different technological initiatives in the agriculture sector and many other sectors, the paper recommends the implementation of frameworks for data protection and sector-specific regulations, along with advocating for the adoption of global standards.

19.6   CONCLUSION

Agriculture is a foundational element of human development, providing sustenance, supporting population growth, and influencing the course of civilization. As we navigate the challenges of the present and future, sustainable and innovative agricultural practices are crucial for ensuring food security and environmental conservation. Sustainable agriculture practices aim to balance the need for food production with environmental conservation, promoting long-term resilience. Balancing technological innovation with ethical and social considerations is pivotal in realizing the full potential of ML in sustainable farming while promoting inclusivity and responsible use of technology. Regular evaluation and adaptation of practices will be essential to address emerging ethical challenges and ensure a positive impact on society and the environment.

By addressing the ethical considerations and social implications discussed in this chapter, the integration of ML technologies in farming can be guided by principles that prioritize fairness, inclusivity, and sustainability. Addressing the challenges requires a collective effort involving governments, agricultural organizations, technology providers, and the farming community. Policymakers, industry stakeholders, and communities must collaborate to develop frameworks that promote the responsible and ethical use of these technologies in agriculture. Policymakers and stakeholders need to work together to create an environment that encourages the adoption of innovative and sustainable agricultural practices while addressing the concerns and limitations faced by farmers.

REFERENCES

[1] Senthil Kumar Swami Durai and Mary Divya Shamili. Smart farming using machine learning and deep learning techniques. *Decision Analytics Journal*, 3:100041, 2022.

[2] Robert Bowen and Wyn Morris. The digital divide: Implications for agribusiness and entrepreneurship. lessons from wales. *Journal of Rural Studies*, 72:75–84, 2019.

[3] Kang, Y. Development of large-scale farming based on explainable machine learning for a sustainable rural economy: the case of cyber risk analysis to prevent costly data breaches. *Applied Artificial Intelligence*, 37(1), 2023. https://doi.org/10.1080/08839 514.2023.2223862

[4] Sunil Kumar, and Marwa M. Eid. A smart solution for sustainable cotton farming: A machine learning approach for visual recognition of leaf diseases. *Full Length Article*, 3(2): 38–8, 2023.

[5] S. M. Banerjee, Shubhash Chandra Chakraborty, and Abhoy Chand Mondal. Machine learning based crop prediction on region wise weather data. *International Journal on Recent and Innovation Trends in Computing and Communication*, 11(1): 145–153, 2023.

[6] Georgios Giannarakis, Vasileios Sitokonstantinou, Roxanne Suzette Lorilla, and Charalampos Kontoes. Personalizing sustainable agriculture with causal machine learning. *arXiv preprint arXiv:2211.03179*, 2022.

[7] Kassim Kalinaki, Wasswa Shafik, Tar J. L. Gutu, and Owais Ahmed Malik. Computer vision and machine learning for smart farming and agriculture practices. In *Artificial Intelligence Tools and Technologies for Smart Farming and Agriculture Practices*, pp. 79–100. IGI Global, 2023.

[8] P. Prema, A. Veeramani, and T. Sivakumar. Machine learning applications in agriculture. *Journal of Agriculture Research and Technology*, 47:126–129, 2022.

[9] Rijwan Khan, Mohammad Ayoub Khan, Mohammad Aslam Ansari, Niharika Dhingra, and Neha Bhati. Machine learning-based agriculture. In *Application of Machine Learning in Agriculture*, pp. 3–27. Academic Press, 2022.

[10] Arthur Carlos, Xayaphone Salinthone, Michelle Nguyen, and Aimee Jacobs. Sustainable agriculture through data analytics. *CSU Journal of Sustainability and Climate Change*, 1(1):2, 2021.

[11] Amit Sharma, Ashutosh Sharma, Alexey Tselykh, Alexander Bozhenyuk, Tanupriya Choudhury, Madani Abdu Alomar, and Manuel S´anchez- Chero. Artificial intelligence and Internet of Things oriented sustainable precision farming: Towards modern agriculture. *Open Life Sciences*, 18:20220713, 2023.

[12] Mohammad Amiri-Zarandi, Rozita A. Dara, Emily Duncan, and Evan D. G. Fraser. Big data privacy in smart farming: A review. *Sustainability*, 14(15), 2022.

[13] K. El Emam and F. K. Dankar. Protecting privacy using k-anonymity. *Journal of the American Medical Informatics Association*, 15:627–637, 2008.

[14] Latanya Sweeney. k-Anonymity: A model for protecting privacy. *International Journal on Uncertainty, Fuzziness and Knowledge-Based Systems*, 10(5):557–570, 2002.

[15] Ninghui Li, Tiancheng Li, and Suresh Venkatasubramanian. t-Closeness: Privacy beyond k-anonymity and l-diversity. In *2007 IEEE 23rd international conference on data engineering* (pp. 106–115). IEEE, 2007.

[16] Renuga Kanagavelu, Qingsong Wei, Zengxiang Li, Haibin Zhang, Juniarto Samsudin, Yechao Yang, Rick Siow Mong Goh, and Shangguang Wang. CE-Fed: Communication efficient multi-party computation enabled federated learning. *Array*, 15:100207, 2022.

[17] Priyank Jain, Manasi Gyanchandani, and Nilay Khare. Differential privacy: Its technological prescriptive using big data. *Journal of Big Data*, 15, 2018.

[18] Yanling Wang, Qian Wang, Lingchen Zhao, and Cong Wang. Differential privacy in deep learning: Privacy and beyond. *Future Generation Computer Systems*, 148:408–424, 2023.

[19] Daniel Morales, Isaac Agudo, and Javier Lopez. Private set intersection: A systematic literature review. *Computer Science Review*, 49:100567, 2023.

[20] Nazish Khalid, Adnan Qayyum, Muhammad Bilal, Ala Al-Fuqaha, and Junaid Qadir. Privacy-preserving artificial intelligence in healthcare: Techniques and applications. *Computers in Biology and Medicine*, 158:106848, 2023.

[21] Kundan Munjal and Rekha Bhatia. A systematic review of homomorphic encryption and its contributions in healthcare industry. *Complex and Intelligent Systems*, 9:3759–3786, 2023.

[22] Rozita Dara, Seyed Mehdi Hazrati Fard, and Jasmin Kaur. Recommendations for ethical and responsible use of artificial intelligence in digital agriculture. *Frontiers in Artificial Intelligence*, 5: 884192, 2022.

[23] Louise Manning, Steve Brewer, Peter J. Craigon, Jeremy Frey, Anabel Gutierrez, Naomi Jacobs, Samantha Kanza, Samuel Munday, Justin Sacks, and Simon Pearson. Artificial intelligence and ethics within the food sector: Developing a common language for technology adoption across the supply chain. *Trends in Food Science & Technology*, 125:33–42, 2022.

[24] Zia Mehrabi, Mollie J. McDowell, Vincent Ricciardi, Christian Levers, Juan Diego Martinez, Natascha Mehrabi, Hannah Wittman, Navin Ramankutty, and Andy Jarvis. The global divide in data-driven farming. *Nature Sustainability*, 4:154–160, 2021.

[25] Madhu Choudhary, P. C. Sharma, H. S. Jat, Ashim Datta, and M. L. Jat. Conservation agriculture: Factors and drivers of adoption and scalable innovative practices in Indo-Gangetic plains of India – A review. *International Journal of Agricultural Sustainability*, 19(1):40–55, 2021.

[26] Akriti Gupta, Aman Chadha, Vijayshri Tiwari, Arup Varma, and Vijay Pereira. Sustainable training practices: Predicting job satisfaction and employee behavior using machine learning techniques. *Asian Business & Management*, 22(5):1–24, 2023.

[27] Rudra Tiwari. The impact of AI and machine learning on job displacement and employment opportunities. *International Journal of Engineering Technologies and Management Research*, 7(1), 2023.

[28] Norman Peter Reeves, Ahmed Ramadan, Victor Giancarlo Sal Y. Rosas Celi, John William Medendorp, Harun Ar-Rashid, Timothy Joseph Krupnik, Anne Namatsi Lutomia, Julia Maria Bello-Bravo, and Barry Robert Pittendrigh. Machine-supported decision-making to improve agricultural training participation and gender inclusivity. *Plos one*, 18(5):e0281428, 2023.

[29] Shilpa Mangesh Pande, Prem Kumar Ramesh, Anmol Anmol, B. R. Aishwarya, Karuna Rohilla, and Kumar Shaurya. Crop recommender system using machine learning approach. In *2021 5th International Conference on Computing Methodologies and Communication (ICCMC)*, pp. 1066–1071. IEEE, 2021.

[30] Damilola Tobiloba Adereti, Maaz Gardezi, Tong Wang, and John Mc-Maine. Understanding farmers' engagement and barrier to machinelearning-based intelligent agricultural decision support systems. *Agronomy Journal*, 116(3):1237–1249, 2023.

[31] Veena Ghuriani, Jyotsna Talreja Wassan, Pragya Deolal, Vidushi Sharma, Dimpy Dalal, and Aditi Goyal. An integrative approach towards recommending farming solutions for sustainable agriculture. *Journal of Experimental Biology and Agricultural Sciences*, 11(2): 306–315, 2023.

[32] Maaz Gardezi, Damilola Tobiloba Adereti, Ryan Stock, and Ayorinde Ogunyiola. In pursuit of responsible innovation for precision agriculture technologies. *Journal of Responsible Innovation*, 9(2):224–247, 2022.

[33] Yuelin Kang. Development of large-scale farming based on explainable machine learning for a sustainable rural economy: The case of cyber risk analysis to prevent costly data breaches. *Applied Artificial Intelligence*, 37(1):2223862, 2023.

[34] Vasileios Sitokonstantinou. Big earth data and machine learning for sustainable and resilient agriculture. *arXiv preprint arXiv:2211.12584*, 2022.

[35] Chimango Nyasulu, Awa Diattara, Assitan Traore, Abdoulaye Deme, and Cheikh Ba. Towards resilient agriculture to hostile climate change in the Sahel region: A case study of machine learning-based weather prediction in Senegal. *Agriculture*, 12(9):1473, 2022.

[36] Jose M. Cadenas, M. Carmen Garrido, and Raquel Mart´ınez-Espa˜na. A methodology based on machine learning and soft computing to design more sustainable agriculture systems. *Sensors*, 23(6):3038, 2023.

[37] Ferne Edwards, Maximilian Manderscheid, and Susan Parham. Terms of engagement: Mobilising citizens in edible nature-based solutions. *Journal of Urbanism: International Research on Placemaking and Urban Sustainability*, 1–22, 2023. https://doi.org/10.1080/17549175.2023.2218356

[38] Rudra Tiwari. Ethical and societal implications of AI and machine learning. *International Journal of Scientific Research in Engineering and Management*, 7(01), 2023.

[39] FAO. Regulatory Framework for Agricultural Data in the Near East and North Africa Region, 2023.

[40] NITI Aayog. National Strategy for Artificial Intelligence AIFORALL. www.niti.gov.in/sites/default/files/2023-03/National-Strategy-for-Artificial-Intelligence.pdf, 2018 (accessed January 16, 2023).

Index

Printed in the United States
by Baker & Taylor Publisher Services